ACCELERATOR PHYSICS

Example Problems with Solutions

T0222469

ACCELERATOR PHYSICS

Example Problems with Solutions

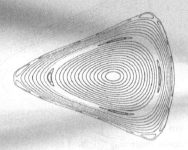

William W MacKay
Brookhaven National Laboratory, USA

Mario Conte
INFN Sezione di Genova, Italy

World Scientific

NEW JERSEY · LONDON · SINGAPORE · BEIJING · SHANGHAI · HONG KONG · TAIPEI · CHENNAI

Published by

World Scientific Publishing Co. Pte. Ltd.

5 Toh Tuck Link, Singapore 596224

USA office: 27 Warren Street, Suite 401-402, Hackensack, NJ 07601

UK office: 57 Shelton Street, Covent Garden, London WC2H 9HE

British Library Cataloguing-in-Publication Data
A catalogue record for this book is available from the British Library.

ISBN-13 978-981-4295-99-4 (pbk)
ISBN-10 981-4295-99-X (pbk)

Printed in Singapore.

Dedication:

We would like to dedicate this book to Johannes Gutenberg and subsequent innovators of replicative printing technology.

Today's mass proliferation of typœs is truly astounding.

The drouping Night thus creepeth on them fast,
  And the sad humour loading their eye liddes,
  As messenger of *Morpheus* on them cast
  Sweet slombring deaw, the which to sleepe them biddes.
  Unto their lodgings then his guestes he riddes:
  Where when all drownd in deadly sleep he findes,
  He to his study goes, and there amiddes
  His Magick bookes and artes of sundry kindes,
He seeks out mighty charmes, to trouble sleepy mindes.

E. Spencer, "The Fairie Queene",
Book I, Canto I, Stanza 36 [63]

Preface

While this book is primarily a solution guide for our previous text, *An Introduction to the Physics of Particle Accelerators* (2nd *Ed.*)[20], we have added some extra commentary and new material in a few places. Given that the problems from CM^1 are written out along with any referenced equations being explicitly quoted, this book may be used as a stand alone volume.

Writing a solution guide for the problems has its pluses and minuses. One drawback with publishing a solution guide is that of coming up with a new set of problems for the next time you teach the course. However, it can make one rethink a problem with perhaps an improved wording or different way of solving the problem. Even on revisiting these problems, the authors can admit to having learned new things while solidifying concepts. In fact every time we teach the course and interact with students, we learn new concepts and ways of presenting the material. As with most problems, there are other possible solutions than those we give.

Most of the problems in the first dozen chapters of CM were actually composed for the first edition[19] which was published two decades ago. We will admit that some of the problems were not tested by assigning them to students, particularly since projects were sometimes assigned during the second half of a course given at Texas A&M University. While a few of the problems may not be perfectly clear, requiring some interpretation, this is not necessarily a bad thing in a graduate physics course.

One problem (CM: 10–2) in particular was originally and incorrectly "solved", and due to sloppy algebra on a blackboard, we gave an incorrect hint. We have provided a corrected solution for this problem along with a

---

[1]Throughout the book we make references to our second edition[20] via prefixing a "CM:" before equation numbers and so forth.

simulation to give an example of how a sextupole may produce a quarter-integer resonance.

The single problem in chapter CM: 12 was written as a research suggestion and we do not give a solution here (You have to win your own prize.); however we do mention a couple of intriguing possibilities which are currently being pursued by various research groups.

We have also included a few comments about various problems with some extra material in chapter 5 on the Lie group behavior of Twiss parameters and in chapter 7 on synchrobetatron coupling and longitudinal canonical variables.

For a few of the problems we have tried to use various computer codes written using free and open source software, e. g. GNU C compiler[37], Octave[24], gnuplot[70], and Maxima[2], which were run on computers using the Ubuntu Linux 10.04[1] operating system. In addition figures have been prepared using the PostScript language[4].

We would like to thank all those students whose *sleepy mindes*[63] we have troubled over the years. MacKay would in particular like to thank Todd Satogata who proofread some of this work while lecturing with MacKay at the U. S. Particle Accelerator School. MacKay would also like to thank some teaching assistants: Ray Fliller, Rama Calaga, and Ilkyoung Shin.

Waldo MacKay
Asheville, NC

Contents

Chapter 1

Problems of Chapter 1: Introduction

1.1 Problem 1–1: Luminosity of Gaussian bunches

a) If the bunches can be described by Gaussian ellipsoids with

$$\rho \propto \exp\left(-\left(\frac{x^2}{2\sigma_x^2} + \frac{y^2}{2\sigma_y^2} + \frac{z^2}{2\sigma_z^2}\right)\right), \tag{1.1}$$

show that the luminosity reduces to

$$\mathcal{L} = f_0 N_b \frac{N_+ N_-}{\pi(4\sigma_x\sigma_y)}, \tag{1.2}$$

where it is assumed that the beams move along the z-axis. The number of particles per bunch in the electron and positron beams are N_- and N_+, respectively. The number of bunches in each beam is given by N_b. (Assume the bunches do not change shape either due to the accelerator optics or the interaction of the beams passing through each other.) b) An e^+e^- storage ring (CESR) operates at 5.3 GeV with 7 bunches of e^+ and 7 bunches of e^- orbiting in opposite directions. Assume the current per bunch is initially 8 mA, and the ring circumference is 768 m. For $\sigma_x = 8.4 \times 10^{-4}$ m, $\sigma_y = 3.5 \times 10^{-5}$ m, and $\sigma_z = 2.2$ cm, what is the initial luminosity in one of the experiments? What is the integrated luminosity of this experiment for a 3 hr run if the beam lifetimes are both 2 hr? (Assume that the beam currents decay exponentially.)

Solution:

a) First we need to normalize the Gaussian distribution. For a single variable x, a Gaussian density function is normalized such that

$$\frac{1}{\sqrt{2\pi}\sigma_x} \int_{-\infty}^{\infty} e^{-\frac{x^2}{2\sigma_x}} dx = 1. \tag{1.3}$$

1

So the density function for the "$\pm$" beam must be

$$\rho_{\pm} = \frac{N_{\pm}}{(2\pi)^{3/2}\sigma_x\sigma_y\sigma_z} \exp\left(-\frac{1}{2}\left(\frac{x^2}{\sigma_x^2} + \frac{y^2}{\sigma_y^2} + \frac{z^2}{\sigma_z^2}\right)\right). \quad (1.4)$$

The luminosity for a single bunch crossing is from Eq. (CM: 1.4) with a slight modification:

$$\mathcal{L}_{\infty} = |\vec{v}_+ - \vec{v}_-| \iiiint \rho_+(x, y, z - ct)\rho_-(x, y, z + ct) \, dx \, dy \, dz \, dt$$

$$= \frac{2cN_+N_-}{(2\pi)^3\sigma_x^2\sigma_y^2\sigma_z^2} \iiiint e^{-x^2/\sigma_x^2} e^{-y^2/\sigma_y^2}$$

$$\times e^{-\frac{1}{2}[(z-ct)^2+(z+ct)^2]/\sigma_z^2} \, dx \, dy \, dz \, dt$$

$$= \frac{N_+N_-}{4\pi^3\sigma_x^2\sigma_y^2\sigma_z^2} \iiiint e^{-x^2/\sigma_x^2} e^{-y^2/\sigma_y^2} e^{-z^2/\sigma_z^2} e^{-\xi^2/\sigma_z^2} \, dx \, dy \, dz \, d\xi, \quad (1.5)$$

with $\xi = ct$ where we have assumed that the velocity of each beam is essentially the speed of light in the high energy limit. Remembering our normalization, the integrals are all of the form

$$\int_{-\infty}^{\infty} e^{-\frac{\zeta^2}{\sigma^2}} \, d\zeta = \sqrt{\pi}\sigma. \quad (1.6)$$

So we must have

$$\mathcal{L}_{\infty} = \frac{N_+N_-}{4\pi\sigma_x\sigma_y}, \quad (1.7)$$

and for N_b bunches per revolution with a revolution frequency f_0, the instantaneous luminosity for one experiment must be

$$\mathcal{L} = f_0 N_b \mathcal{L}_{\infty} = f_0 N_b \frac{N_+N_-}{4\pi\sigma_x\sigma_y}. \quad (1.8)$$

b) For CESR with the given parameters,

$$f_0 = \frac{3 \times 10^8 \text{ m/s}}{768 \text{ m}} = 3.91 \times 10^5 \text{ Hz}, \quad (1.9)$$

$$N_{\pm} = \frac{I_0}{ef_0} = \frac{8 \text{ mA}}{(1.6 \times 10^{-19} \text{ C}) \times (3.9 \times 10^5 \text{ s}^{-1})} = 1.28 \times 10^{11}, \quad (1.10)$$

so

$$\mathcal{L}_0 = \frac{(3.91 \times 10^5 \text{Hz}) \times 7 \times (1.28 \times 10^{11})^2}{4\pi \times (8.4 \times 10^{-4} \text{ m}) \times (3.5 \times 10^{-5} \text{ m})} \simeq 1.21 \times 10^{-31} \text{ cm}^{-2}\text{s}^{-1}$$

$$(1.11)$$

for the initial luminosity of one experiment. If we assume that each beam decays exponentially, then $I_{\pm}(t) = I_0 e^{-t/\tau}$ with $\tau = 2$ hr. Since $\mathcal{L} \propto$

$I_+(t)I_-(t) \propto e^{-2t/\tau}$, we find the total integrated luminosity over a 3 hr run would be

$$L = \mathcal{L}_0 \int_0^{3\text{hr}} e^{-\frac{2t}{\tau}} dt = \mathcal{L}_0 \frac{\tau}{2} \left[1 - e^{-2(3\text{hr})/\tau} \right]$$

$$= (1.2 \times 10^{-31} \text{cm}^{-2}\text{s}^{-1}) \times \left(\frac{3 \times 3600}{2} \text{ s} \right) \times (1 - e^{-3}) = 6.2 \times 10^{34} \text{ cm}^{-2}$$

$$= 62 \text{ nb}^{-1}, \quad \text{where } 1 \text{ b} = 10^{-24} \text{cm}^{-2}. \tag{1.12}$$

1.2   Problem 1–2: Brightness of laser beam

Calculate the brightness of a NdYAG laser with the following parameters:

$$\lambda = 1.064\mu \text{ m} \tag{1.13}$$

$$\text{Power} = 20 \text{ W} \tag{1.14}$$

$$\text{Bandwidth } \frac{\Delta\omega}{2\pi} = 120 \text{ GHz} \tag{1.15}$$

$$\text{Beam divergence} = 10 \text{ mrad.} \tag{1.16}$$

Solution:

The energy of a single photon is

$$u = 2\pi\frac{\hbar c}{\lambda} = \frac{2\pi \times (197 \text{ MeV fm})}{1.064 \times 10^{-6} \text{ m}} = 1.165 \text{ eV}, \tag{1.17}$$

so the flux of photons is $dn/dt = P/u = 1.071 \times 10^{20} \text{ s}^{-1}$. In laser physics, the divergence angle θ_d is generally specifies a narrow beam with an opening cone of $\pm\theta_d/2$ and in this case yields a solid angle of

$$d\Omega = \int_0^{2\pi} \int_0^{\theta_d/2} \sin\theta \, d\theta \, d\phi \simeq \pi \times (0.005)^2 = 1.5708 \times 10^{-4} \text{ sr.} \tag{1.18}$$

The bandwidth of 120 GHz can be converted into wavelength units by

$$\frac{|\Delta\lambda|}{\lambda} = \frac{|\Delta\omega|}{\omega} = \frac{|\Delta u|}{u}, \tag{1.19}$$

and

$$\begin{aligned}
\Delta u &= h\Delta\nu = \hbar\Delta\omega \\
&= (6.582 \times 10^{-16} \text{ eVs}) \times 2\pi \times (1.2 \times 10^{11} \text{ Hz}) \\
&= 4.963 \times 10^{-4} \text{ eV.}
\end{aligned} \tag{1.20}$$

Using Eq. (CM 1.5) we then get

$$\begin{aligned}
d\Phi_\Omega &= 1000\frac{d^4n}{dt \, d\Omega \, (|du|/u)} = \frac{1000 \times (1.071 \times 10^{20} \text{ s})}{(1.5708 \times 10^{-4}) \times (4.963 \times 10^{-4}/1.165)} \\
&= 1.601 \times 10^{30} \text{ sr}^{-1}\text{s}^{-1}.
\end{aligned} \tag{1.21}$$

1.3 Problem 1–3: Equations of motion

In a fixed Cartesian coordinates system, show that the equations of motion for a charged particle moving in a magnetic field may be written as

$$x'' = \frac{q}{p}(1 + x'^2 + y'^2)^{\frac{1}{2}}[y'B_z - (1 + x'^2)B_y + x'y'B_x], \quad \text{and} \quad (1.22)$$

$$y'' = -\frac{q}{p}(1 + x'^2 + y'^2)^{\frac{1}{2}}[x'B_z - (1 + y'^2)B_x + x'y'B_y], \quad (1.23)$$

where the primes denote derivatives with respect to z (i. e., $x' = dx/dz$). Here it has been assumed that the electric field is zero and that $dz/dt \neq 0$.

Solution[64]:

In these conditions, the motion equation is

$$q(\vec{v} \times \vec{B}) = \frac{d\vec{p}}{dt} = \frac{d}{dt}(\gamma m \vec{v}) = \gamma m \frac{d\vec{v}}{dt} + m \frac{d\gamma}{dt}\vec{v} \qquad (1.24)$$

with $\gamma = \left[1 - \frac{v^2}{c^2}\right]^{-\frac{1}{2}}$, $\vec{v} = \hat{x}\frac{dx}{dt} + \hat{y}\frac{dy}{dt} + \hat{z}\frac{dz}{dt}$ and $\vec{B} = \hat{x}B_x + \hat{y}B_y + \hat{z}B_z$. Due to the trivial relation

$$\frac{d}{dt} = \frac{dz}{dt}\frac{d}{dz} = \dot{z}\frac{d}{dz} \qquad (1.25)$$

we can write

$$\vec{v} = \dot{z}(\hat{x}x' + \hat{y}y' + \hat{z} \cdot 1) \quad \text{or} \quad v = |\vec{v}| = \dot{z}\sqrt{1 + x'^2 + y'^2} \qquad (1.26)$$

where the prime represents the derivative with respect to z. The vector product in the left side of Eq. (1.24) becomes

$$\vec{v} \times \vec{B} = \begin{vmatrix} \hat{x} & \hat{y} & \hat{z} \\ \dot{z}x' & \dot{z}y' & \dot{z} \\ B_x & B_y & B_z \end{vmatrix} = \begin{pmatrix} \dot{z}(y'B_z - B_y) \\ \dot{z}(B_x - x'B_z) \\ \dot{z}(x'B_y - y'B_x) \end{pmatrix} \qquad (1.27)$$

Besides, from Eq. (1.26) we have

$$\frac{d\vec{v}}{dt} = \dot{z}^2(\hat{x}x'' + \hat{y}y'') \quad \text{and} \quad \frac{dv}{dt} = \dot{z}^2 \frac{x'x'' + y'y''}{\sqrt{1 + x'^2 + y'^2}} \qquad (1.28)$$

Therefore the components of Eq. (1.24) can be written as

$$\gamma m \dot{z}^2 x'' + m \frac{d\gamma}{dt} \dot{z} x' = q \dot{z}(y' B_z - B_y) \tag{1.29}$$

$$\gamma m \dot{z}^2 y'' + m \frac{d\gamma}{dt} \dot{z} y' = q \dot{z}(B_x - x' B_z) \tag{1.30}$$

$$m \frac{d\gamma}{dt} \dot{z} = q \dot{z}(x' B_y - y' B_x) \tag{1.31}$$

or

$$\gamma m \dot{z} x'' + \frac{m}{c^2} \dot{z}^3 \gamma^3 (x' x'' + y' y'') x' = q(y' B_z - B_y) \tag{1.32}$$

$$\gamma m \dot{z} y'' + \frac{m}{c^2} \dot{z}^3 \gamma^3 (x' x'' + y' y'') y' = q(B_x - x' B_z) \tag{1.33}$$

$$\frac{m}{c^2} \dot{z}^3 \gamma^3 (x' x'' + y' y'') = q(x' B_y - y' B_x) \tag{1.34}$$

having considered that

$$\frac{d\gamma}{dt} = \gamma^3 \frac{v}{c^2} \frac{d\vec{v}}{dt} = \frac{\dot{z}^3 \gamma^3}{c^2} (x' x'' + y' y'') \tag{1.35}$$

Inserting the Eq. 1.34 into Eqs. (1.32 and 1.33) we obtain

$$\gamma m \dot{z} x'' + q(x' B_y - y' B_x) x' = q(y' B_z - B_y) \tag{1.36}$$

$$\gamma m \dot{z} y'' + q(x' B_y - y' B_x) y' = q(B_x - x' B_z) \tag{1.37}$$

which will yield

$$x'' = \frac{q}{\gamma m \dot{z}} [y' B_z - (1 + x'^2) B_y + x' y' B_x] \tag{1.38}$$

$$y'' = -\frac{q}{\gamma m \dot{z}} [x' B_z - (1 + x'^2) B_x + x' y' B_y] \tag{1.39}$$

or

$$x'' = \frac{q}{p} \sqrt{1 + x'^2 + y'^2} [y' B_z - (1 + x'^2) B_y + x' y' B_x] \tag{1.40}$$

$$y'' = -\frac{q}{p} \sqrt{1 + x'^2 + y'^2} [x' B_z - (1 + x'^2) B_x + x' y' B_y] \tag{1.41}$$

having considered Eq. (1.26) and that $p = |\vec{p}| = \gamma m v$.

1.4 Problem 1–4: Decay of a pion to a muon

Consider a charged pion decaying into a muon plus an antineutrino:

$$\pi^- \rightarrow \mu^- + \bar{\nu}_\mu. \tag{1.42}$$

Use $M_{\pi\pm} = 140 \text{ eV}/c^2$, $m_\mu = 106 \text{ MeV}/c^2$, and $m_{\bar{\nu}} = 0$.

a) In the rest system of the pion, what are the energies and momenta of the muon and antineutrino?
b) For a moving pion with total energy $U_\pi = \gamma M_\pi c^2$ find an expression for the direction, θ_μ of the muon relative to the pion in the lab in terms of the angle θ_μ^* in the in the pion's rest system.

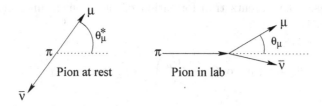

Pion at rest   Pion in lab

Solution:

In the pion rest frame the total energy is just the rest-mass energy which, after the decay, transforms into the sum of the muon and neutrino total energies; namely

$$M_\pi c^2 = U_\mu^* + U_{\bar{\nu}}^* = m_\mu c^2 + W_\mu^* + W_{\bar{\nu}}^* \tag{1.43}$$

since $m_{\bar{\nu}}^* = 0$. Besides, the pion momentum in its rest frame is obviously zero and this implies that

$$\vec{p}_\mu^* + \vec{p}_{\bar{\nu}}^* = 0 \quad \Rightarrow \quad \vec{p}_\mu^* = -\vec{p}_{\bar{\nu}}^* \quad \Rightarrow \quad |\vec{p}_\mu^*| = |\vec{p}_{\bar{\nu}}^*| \quad \Rightarrow \quad p_\mu^* = p_{\bar{\nu}}^* \tag{1.44}$$

or

$$p_\mu^* c = \sqrt{(U_\mu^*)^2 - (m_\mu c^2)^2} = p_{\bar{\nu}}^* c = W_{\bar{\nu}}^* \tag{1.45}$$

which combined with Eq. (1.43) gives rise to

$$M_\pi c^2 - U_\mu^* = \sqrt{(U_\mu^*)^2 - (m_\mu c^2)^2}, \quad \text{or} \quad (1.46)$$

$$U_\mu^* = \frac{(M_\pi c^2)^2 + (m_\mu c^2)^2}{2(M_\pi c^2)}. \tag{1.47}$$

Hence we have

$$W_\mu^* = \frac{(M_\pi c^2)^2 + (m_\mu c^2)^2}{2(M_\pi c^2)} - m_\mu c^2,$$

$$= \frac{(M_\pi c^2 - m_\mu c^2)^2}{2(M_\pi c^2)} = 29.87 \text{ MeV} \tag{1.48}$$

$$W_{\bar\nu}^* = M_\pi c^2 - m_\mu c^2 - W_\mu^* = 4.13 \text{ MeV}. \tag{1.49}$$

In this case, the Lorentz transformation of the momentum longitudinal component is

$$p_\mu \cos\theta_\mu = \gamma\left(p_\mu^* \cos\theta_\mu^* + \beta\frac{U_\mu^*}{c}\right) = \gamma p_\mu^* \cos\theta_\mu^* + \beta\gamma\frac{U_\mu^*}{c}, \tag{1.50}$$

where

$$\gamma = \frac{U_\pi}{M_\pi c^2}, \quad \text{and} \quad \beta\gamma = \sqrt{\gamma^2 - 1} = \frac{U_\pi^2 - (M_\pi c^2)^2}{M_\pi c^2} = \frac{p_\pi}{M_\pi c}, \tag{1.51}$$

which altogether yield

$$\cos\theta_\mu = \frac{U_\pi p_\mu^* \cos\theta_\mu^* + U_\mu^* p_\pi}{p_\mu M_\pi c^2}. \tag{1.52}$$

1.5   Problem 1–5: Collider vs fixed-target energies

The Tevatron collides protons ($m_p = 0.938$ GeV) at 1 TeV per beam. What is the equivalent proton beam energy required to produce the same center-of-mass energy with a stationary hydrogen target? How fast would you have to drive your new 1.3 ton VW Beetle to have the same kinetic energy as a bunch of 10^{13} protons with this energy? (The speed of sound in air is 330 m/s.)

Solution:

In symmetric colliders the center of mass is standing, meaning that laboratory and center-of-mass frames coincide; namely

$$U_{\text{CM}} = 2\,U_{\text{Tevatron}} = 2\,U_{\text{T}} = 2 \text{ TeV}. \tag{1.53}$$

Then consider the invariant

$$p^\mu p_\mu c^2 = U^2 - \vec{p}^{\,2}c^2 = \frac{1}{c^2}(U_{\text{CM}})^2 - P^2_{\text{CM}} = U^2_{\text{CM}}, \tag{1.54}$$

or for the beam (x) hitting a fixed target:

$$p^\mu p_\mu c^2 = (U_x + m_p c^2)^2 - p_x^2 c^2 = (U_x + m_p c^2)^2 - (U_x^2 - m_p^2 c^4)$$
$$\simeq 2U_x m_p c^2, \tag{1.55}$$

since $U_x \gg m_p c^2$. Solving for U_x yields

$$U_x = \frac{2\,U_{\text{T}}^2}{m_p c^2} = \frac{2 \times 10^6}{0.938} = 2.13 \times 10^6 \text{ GeV}. \tag{1.56}$$

Assuming that the VW Beetle is nonrelativistic, it would have a kinetic energy of

$$W_{\text{VW}} = \frac{1}{2} M_{\text{VW}} v^2 = N_{\text{Bunches}} U_x$$
$$= 10^{13} \times 2.13 \times 10^{15} [\text{eV}] \times 1.6 \times 10^{-19} [\text{J/eV}]$$
$$= 3.41 \times 10^9 \text{ J}, \tag{1.57}$$

or

$$v = \sqrt{\frac{2W_{\text{VW}}}{M_{\text{VW}}}} = 2290 \text{ m/s} \sim \text{Mach 7}. \tag{1.58}$$

1.6 Problem 1–6: HERA center-of-mass system

HERA collides 920 GeV protons with 27.5 GeV electrons with zero crossing angle.

a) What is the center-of-mass energy?
b) What is the velocity of the center of mass in the lab system?

Solution:

Let us recall the expression of the momentum energy four-vector

$$p^\mu = \left(\frac{U}{c}, \vec{p} \right), \quad \text{with} \quad U = \sqrt{(mc^2)^2 + (pc)^2}, \tag{1.59}$$

whose squared modulus is an invariant. In fact:

$$p^\mu p_\mu = \frac{U^2}{c^2} - (p_x^2 + p_y^2 + p_z^2) = \frac{m^2 c^4}{c^2} + \frac{p^2 c^2}{c^2} - p^2 = (mc)^2. \tag{1.60}$$

Therefore the following identities subsist

$$U^2 - (pc)^2 = U'^2 - (p'c)^2 = U^{*\,2}. \tag{1.61}$$

Where the asterisk "$*$", as well as the subscript "cm", indicates the center-of-mass frame, whose momentum is obviously null. Labeling with L the laboratory frame, we have

$$U_{\text{cm}}^2 = U_{\text{L}}^2 - (p_{\text{L}} c)^2, \tag{1.62}$$

with

$$U_{\text{L}} = U_{\text{p}} + U_{\text{e}}, \quad \text{and} \quad p_{\text{L}} = |\vec{p}_{\text{p}} - \vec{p}_{\text{e}}|, \tag{1.63}$$

where $U_{\text{p}} = 920$ GeV and $U_{\text{e}} = 27.5$ GeV are the given data of the problem. Moreover, the momenta and energies coincide since we are in an ultra-relativistic regime. Hence the center-of-mass energy results in

$$U_{\text{cm}} = \sqrt{(U_{\text{p}} + U_{\text{e}})^2 + |\vec{p}_{\text{p}} - \vec{p}_{\text{e}}|^2} = 318 \text{ GeV}. \tag{1.64}$$

As far as the center-of-mass velocity is concerned, we simply have

$$v_{\text{cm}} = \beta_{\text{cm}} c = \frac{p_{\text{L}}}{U_{\text{L}}} c = 0.889\, c = 2.67 \times 10^8 \text{ ms}^{-1}. \tag{1.65}$$

1.7 Problem 1–7: SLAC rf voltage and power

The Stanford Linear Accelerator is 3.05 km long and can accelerate electrons up to 50 GeV.

a) What is the average accelerating gradient of the rf cavities?
b) For bunches of 4×10^{10} electrons per bunch and a duty cycle of 100 Hz, what is the power transferred to the beam?

Solution:

a) The average accelerating gradient is

$$G = \frac{50 \text{ GeV}}{3.05 \text{ km}} = 16.4 \text{ MeV/m}. \tag{1.66}$$

b) The power transferred to the beam is given by the duty cycle f times the total energy $N_e U_e$ of the electron beam:

$$P = 100 \text{ [Hz]} \times 4 \times 10^{10} \times 50 \times 10^{9} \text{ [eV]} \times 1.6 \times 10^{-19} \text{ [C/e]}$$
$$= 32 \text{ kW}. \tag{1.67}$$

1.8   Problem 1–8: Fixed-target interaction rate

An experiment has a 10 cm long liquid hydrogen target with a density of

$$\rho = .063 \text{ g/cm}^3. \tag{1.68}$$

Estimate the interaction rate for p+p collisions for a beam of 10^{13} protons every two minutes. Assume the total cross section is 40 mb. Note: 1 barn = 10^{-24} cm^2 and Avogadro's number is 6×10^{23}.

Solution:

Assuming that the diameter of the beam is smaller than the radius of the target so that the whole beam passes through the target:

$$
\begin{aligned}
\frac{dN}{dt} &= 0.063[\text{g/cm}^3] \times 10[\text{cm}] \times \frac{6 \times 10^{23}}{1[\text{g}]} \times \frac{10^{13}}{120[\text{s}]} \\
&\quad \times 40 \times 10^{-3}[\text{b}] \times 10^{-24}[\text{cm}^2/\text{b}] \\
&= 1.26 \times 10^9 [\text{s}^{-1}].
\end{aligned}
\tag{1.69}
$$

Chapter 2

Problems of Chapter 2: Equations of Motion for Weak Focusing

2.1 Problem 2–1: Equations of motion with dispersion

Derive Eqs. (CM: 2.39 and CM: 2.40), but keep terms to second order in x, y, and δ.

$$\frac{d^2x}{d\theta^2} + (1 - n)x = \rho\frac{\delta\rho}{p}. \qquad \text{(CM:2.39)}$$

$$\frac{d^2y}{d\theta^2} + ny \simeq 0. \qquad \text{(CM:2.40)}$$

Solution:

Let us rewrite Eq. (CM: 2.5)

$$\frac{d^2x}{d\theta^2} + \left(\frac{qB_y}{\beta\gamma mc}R - 1\right)R = 0, \qquad \text{(CM: 2.5)}$$

as

$$\frac{d^2x}{d\theta^2} + \left(\frac{qB_y}{p}R - 1\right)R = 0, \qquad (2.1)$$

and carry out the substitutions

$$\frac{1}{p} = \frac{1}{p_0 + \delta p} \simeq \frac{1}{p_0}(1 - \delta + \delta^2) \quad \text{with} \quad \delta = \frac{\delta p}{p_0}, \qquad (2.2)$$

$$B_y \simeq B_0 + \frac{\partial B_y}{\partial x}x + \frac{1}{2}\frac{\partial^2 B_y}{\partial x^2}x^2, \qquad (2.3)$$

and

$$R = \rho + x = \rho\left(1 + \frac{x}{\rho}\right). \qquad (2.4)$$

Reiterating then the procedure which lead us from Eq. (CM: 2.5) to Eq. (CM: 2.12)

$$\frac{d^2x}{d\theta^2} + (1-n)x = 0,$$ (CM: 2.12)

we obtain:

$$\frac{d^2x}{d\theta^2} = -\left[\frac{qB_0\rho}{p_0}(1-\delta+\delta^2)\left(1+\frac{1}{B_0}\frac{\partial B_y}{\partial x}x + \frac{1}{2B_0}\frac{\partial^2 B_y}{\partial x^2}x^2 + \delta^2\right)\right.$$
$$\left. \times\left(1+\frac{x}{\rho}\right) - 1\right](\rho+x),$$ (2.5)

and considering that

$$p_0 = qB_0\rho$$ (2.6)

we have:

$$\frac{d^2x}{d\theta^2} = -\left[\left(1+\frac{1}{B_0}\frac{\partial B_y}{\partial x}x + \frac{1}{2B_0}\frac{\partial^2 B_y}{\partial x^2}x^2 - \delta - \frac{1}{B_0}\frac{\partial B_y}{\partial x}x\delta + \delta^2\right)\right.$$
$$\left.\left(1+\frac{x}{\rho}\right)-1\right](\rho+x)$$
$$= -\left(\frac{1}{B_0}\frac{\partial B_y}{\partial x}x + \frac{1}{2B_0}\frac{\partial^2 B_y}{\partial x^2}x^2 - \delta - \frac{1}{B_0}\frac{\partial B_y}{\partial x}x\delta + \delta^2\right.$$
$$\left. +\frac{x}{\rho} + \frac{1}{\rho B_0}\frac{\partial B_y}{\partial x}x^2 - \frac{x\delta}{\rho}\right)(\rho+x),$$ (2.7)

or

$$\frac{d^2x}{d\theta^2} + \left[\left(1+\frac{\rho}{B_0}\frac{\partial B_y}{\partial x}\right)\left(1-\frac{\delta p}{p_0}\right) - \frac{\delta p}{p_0}\right]x$$
$$+\left[\frac{1}{\rho} + \frac{2}{B_0}\frac{\partial B_y}{\partial x} + \frac{\rho}{2B_0}\frac{\partial^2 B_y}{\partial x^2}\right]x^2 = \rho\frac{\delta p}{p_0}\left(1-\frac{\delta p}{p_0}\right),$$ (2.8)

which, considering the field index definition

$$n = -\frac{\rho}{B_0}\left(\frac{\partial B_y}{\partial x}\right)_{x=0},$$ (CM: 2.13)

becomes

$$\frac{d^2x}{d\theta^2} + \left[(1-n)\left(1-\frac{\delta p}{p_0}\right) - \frac{\delta p}{p_0}\right]x + \left[\frac{1}{\rho} - \frac{2n}{\rho} + \frac{\rho}{2B_0}\frac{\partial^2 B_y}{\partial x^2}\right]x^2$$
$$= \rho\frac{\delta p}{p_0}\left(1-\frac{\delta p}{p_0}\right).$$ (2.9)

Rewriting Eq. (CM: 2.6)

$$\frac{d^2y}{d\theta^2} - \frac{qB_x}{\beta\gamma mc}R^2 = 0,\qquad\text{(CM: 2.6)}$$

as

$$\frac{d^2y}{d\theta^2} - \frac{qB_y}{p}R^2 = 0,\qquad(2.10)$$

we consider that, to second order,

$$B_x \simeq \frac{\partial B_x}{\partial y}y + \frac{1}{2}\frac{\partial^2 B_y}{\partial x^2}y^2 = \frac{\partial B_y}{\partial x}y + \frac{1}{2}\frac{\partial^2 B_y}{\partial x^2}y^2 \qquad(2.11)$$

since $B(y = 0) = 0$ (field lines symmetry) and $\frac{\partial B_x}{\partial y} = \frac{\partial B_y}{\partial x}$ (Ampere's law $\nabla \times \vec{B} = 0$). Therefore, taking into account the set of approximations introduced before, Eq. (2.10) becomes

$$
\begin{aligned}
\frac{d^2y}{d\theta^2} &\simeq \frac{q}{p_0}(1 - \delta + \delta^2)\left(\frac{\partial B_y}{\partial x}y + \frac{1}{2}\frac{\partial^2 B_y}{\partial x^2}y^2\right)(\rho^2 + 2\rho x + x^2)\\
&\simeq \frac{q}{p_0}\left(\frac{\partial B_y}{\partial x}y + \frac{1}{2}\frac{\partial^2 B_y}{\partial x^2}y^2 - \frac{\partial B_y}{\partial x}y\delta\right)(\rho^2 + 2\rho x + x^2)\\
&\simeq \frac{q}{p_0}\left(\rho^2\frac{\partial B_y}{\partial x}y + 2\rho\frac{\partial B_y}{\partial x}xy + \frac{\rho^2}{2}\frac{\partial^2 B_y}{\partial x^2}y^2 - \rho^2\frac{\partial B_y}{\partial x}y\delta\right)\\
&\simeq \frac{qB_0\rho}{p_0}\left(\frac{\rho}{B_0}\frac{\partial B_y}{\partial x}y + \frac{2}{B_0}\frac{\partial B_y}{\partial x}xy + \frac{\rho}{2B_0}\frac{\partial^2 B_y}{\partial x^2}y^2 - \frac{\rho}{B_0}\frac{\partial B_y}{\partial x}y\delta\right),
\end{aligned}
$$
$$(2.12)$$

or

$$\frac{d^2y}{d\theta^2} + n\left(1 - \frac{\delta p}{p_0}\right)y + \frac{2n}{\rho}xy - \frac{\rho}{2B_0}\frac{\partial^2 B_y}{\partial x^2}y^2 = 0.\qquad(2.13)$$

2.2 Problem 2–2: Horizontal transport matrix with dispersion

Work out the equivalent for Eq. (CM:2.55) with dispersion.

$$\mathbf{M_H} = \begin{pmatrix} \cos\frac{\sqrt{1-n}\pi}{2} - \frac{l_0}{2\rho}\sqrt{1-n}\sin\frac{\sqrt{1-n}\pi}{2} \\ -\frac{\sqrt{1-n}}{\rho}\sin\frac{\sqrt{1-n}\pi}{2} \end{pmatrix}$$

$$\begin{pmatrix} l_0\cos\frac{\sqrt{1-n}\pi}{2} + \frac{\rho}{\sqrt{1-n}}\left[1 - \frac{l_0^2(1-n)}{4\rho^2}\right]\sin\frac{\sqrt{1-n}\pi}{2} \\ \cos\frac{\sqrt{1-n}\pi}{2} - \frac{l_0}{2\rho}\sqrt{1-n}\sin\frac{\sqrt{1-n}\pi}{2} \end{pmatrix}. \quad \text{(CM:2.55)}$$

Solution:

Let us modify the 2×2-matrix Eq. (CM: 2.54)

$$\mathbf{M_H} = \begin{pmatrix} 1 & \frac{l_0}{2} \\ 0 & 1 \end{pmatrix}\begin{pmatrix} \cos\frac{\sqrt{1-n}\pi}{2} & \frac{\rho}{\sqrt{1-n}}\sin\frac{\sqrt{1-n}\pi}{2} \\ -\frac{\sqrt{1-n}}{\rho}\sin\frac{\sqrt{1-n}\pi}{2} & \cos\frac{\sqrt{1-n}\pi}{2} \end{pmatrix}\begin{pmatrix} 1 & \frac{l_0}{2} \\ 0 & 1 \end{pmatrix}, \quad \text{(CM: 2.54)}$$

in the following way. The free-flight 2×2 matrices are changed into two 3×3 matrices. The bend matrix has to be substituted by the one which appears in Eq. (CM:2.46)

$$\begin{pmatrix} x \\ x' \\ \frac{\delta p}{p} \end{pmatrix} = \begin{pmatrix} \cos\sqrt{1-n}\theta & \frac{\rho}{\sqrt{1-n}}\sin\sqrt{1-n}\theta & \frac{\rho}{1-n}(1-\cos\sqrt{1-n}\theta) \\ -\frac{\sqrt{1-n}}{\rho}\sin\sqrt{1-n}\theta & \cos\sqrt{1-n}\theta & \frac{1}{\sqrt{1-n}}\sin\sqrt{1-n}\theta \\ 0 & 0 & 1 \end{pmatrix}$$

$$\begin{pmatrix} x_0 \\ x'_0 \\ \left(\frac{\delta p}{p}\right)_0 \end{pmatrix},$$

$$\text{(CM: 2.46)}$$

but with $\theta = \frac{1}{2}\pi$:

$$\mathbf{M_{H3}} = \begin{pmatrix} \mathbf{L} & 0 \\ & 0 \\ 0\ 0\ 1 \end{pmatrix}\begin{pmatrix} \mathbf{B} & B_{13} \\ & B_{23} \\ 0\ 0\ 1 \end{pmatrix}\begin{pmatrix} \mathbf{L} & 0 \\ & 0 \\ 0\ 0\ 1 \end{pmatrix} = \begin{pmatrix} \mathbf{LB} & L B_{13} \\ & B_{23} \\ 0\ 0 & 1 \end{pmatrix}\begin{pmatrix} \mathbf{L} & 0 \\ & 0 \\ 0\ 0\ 1 \end{pmatrix}$$

$$= \begin{pmatrix} \mathbf{M_H} & L B_{13} \\ & B_{23} \\ 0\ 0 & 1 \end{pmatrix} = \begin{pmatrix} \mathbf{M_H} & \mathbf{D} \\ 0\ 0 & 1 \end{pmatrix}, \quad (2.14)$$

with the 2×2-block matrix $\mathbf{M_H}$ of Eq. (CM: 2.55) and where

$$\mathbf{L} = \begin{pmatrix} 1 & \frac{l_0}{2} \\ 0 & 1 \end{pmatrix}, \quad \text{and} \quad \mathbf{B} = \begin{pmatrix} \cos \frac{\sqrt{1-n}\pi}{2} & \frac{\rho}{\sqrt{1-n}} \sin \frac{\sqrt{1-n}\pi}{2} \\ -\frac{\sqrt{1-n}}{\rho} \sin \frac{\sqrt{1-n}\pi}{2} & \cos \frac{\sqrt{1-n}\pi}{2} \end{pmatrix}. \quad (2.15)$$

From Eq. (CM: 2.46) with $\theta = \pi/2$ we have:

$$\begin{pmatrix} B_{13} \\ B_{23} \end{pmatrix} = \begin{pmatrix} \frac{\rho}{1-n} \left(1 - \cos \frac{\sqrt{1-n}\pi}{2} \right) \\ \frac{1}{\sqrt{1-n}} \sin \frac{\sqrt{1-n}\pi}{2} \end{pmatrix}. \quad (2.16)$$

So we only need to calculate the dispersion terms of the vector $\mathbf{D}$, since the 2×2-block $\mathbf{M_H}$ remains unchanged. The desired dispersion vector in the third column of the matrix $\mathbf{M_{H3}}$ will be

$$\begin{aligned} \mathbf{D} &= \begin{pmatrix} 1 & \frac{l_0}{2} \\ 0 & 1 \end{pmatrix} \begin{pmatrix} \frac{\rho}{1-n} \left(1 - \cos \frac{\sqrt{1-n}\pi}{2} \right) \\ \frac{1}{\sqrt{1-n}} \sin \frac{\sqrt{1-n}\pi}{2} \end{pmatrix} \\ &= \begin{pmatrix} \frac{\rho}{1-n} \left(1 - \cos \frac{\sqrt{1-n}\pi}{2} \right) + \frac{l_0}{2\sqrt{1-n}} \sin \frac{\sqrt{1-n}\pi}{2} \\ \frac{1}{\sqrt{1-n}} \sin \frac{\sqrt{1-n}\pi}{2} \end{pmatrix}. \quad (2.17) \end{aligned}$$

2.3   Problem 2–3: Betatron's momentum compaction

Calculate the momentum compaction factor, α_p, for a betatron with a field index, n.

Solution:

Recall the definition of momentum compaction:

$$\alpha = \frac{dL}{L} \bigg/ \frac{dp}{p} = \frac{p}{L}\frac{dL}{dp}. \qquad\qquad \text{(CM: 1.39)}$$

Since in the betatron $L = 2\pi\rho$, we have

$$\frac{dL}{L} = \frac{d\rho}{\rho}. \qquad\qquad (2.18)$$

Then recalling Eq. (CM:2.78)

$$\frac{\delta p}{p} = \frac{d\rho}{\rho} + \frac{dB}{B} = \left(1 + \frac{\rho}{B}\frac{\partial B}{\partial \rho}\right)\frac{d\rho}{\rho} = (1-n)\frac{dp}{p}, \qquad \text{(CM: 2.78)}$$

we quickly obtain

$$\alpha = \frac{\frac{d\rho}{\rho}}{(1-n)\frac{d\rho}{\rho}} = \frac{1}{1-n}. \qquad\qquad (2.19)$$

2.4 Problem 2–4: Combining two dipoles

Show by explicit multiplication of matrices that the transfer matrices (both vertical and horizontal) for a combination of two sector magnets with identical bending radii, but with different bend angles θ_1 and θ_2, is equivalent to a sector magnet with a bend of $\theta_1 + \theta_2$, if there is no drift between the two magnets.

Solution:

Let us start the transport matrices from Eqs. (CM: 2.47 and CM: 2.36)

$$\mathbf{M_H} = \begin{pmatrix} \cos\theta & \rho\sin\theta & \rho(1-\cos\theta) \\ -\frac{1}{\rho}\sin\theta & \cos\theta & \sin\theta \\ 0 & 0 & 1 \end{pmatrix}, \qquad \text{(CM: 2.47)}$$

$$\mathbf{M_V} = \begin{pmatrix} 1 & \rho\theta \\ 0 & 1 \end{pmatrix}, \qquad \text{(CM: 2.32)}$$

respectively for the horizontal and vertical matrices.

Let us do the vertical plane first:

$$\mathbf{N_V} = \mathbf{M_V}(\theta_2)\mathbf{M_V}(\theta_1) = \begin{pmatrix} 1 & \rho\theta_2 \\ 0 & 1 \end{pmatrix}\begin{pmatrix} 1 & \rho\theta_1 \\ 0 & 1 \end{pmatrix} = \begin{pmatrix} 1 & \rho(\theta_2+\theta_1) \\ 0 & 1 \end{pmatrix}, \quad (2.20)$$

so the vertical plane certainly works. For the horizontal plane we have:

$$\mathbf{N_H} = \mathbf{M_H}(\theta_2)\mathbf{M_H}(\theta_1) \qquad (2.21)$$

$$= \begin{pmatrix} \cos\theta_2 & \rho\sin\theta_2 & \rho(1-\cos\theta_2) \\ -\frac{1}{\rho}\sin\theta_2 & \cos\theta_2 & \sin\theta_2 \\ 0 & 0 & 1 \end{pmatrix}\begin{pmatrix} \cos\theta_1 & \rho\sin\theta_1 & \rho(1-\cos\theta_1) \\ -\frac{1}{\rho}\sin\theta_1 & \cos\theta_1 & \sin\theta_1 \\ 0 & 0 & 1 \end{pmatrix},$$

For the first row we have:

$$N_{11} = \cos\theta_2\cos\theta_1 - \sin\theta_2\sin\theta_1 = \cos(\theta_2+\theta_1), \qquad (2.22)$$

$$N_{12} = \rho(\cos\theta_2\sin\theta_1 + \sin\theta_2\cos\theta_1) = \rho\sin(\theta_2+\theta_1), \qquad (2.23)$$

$$N_{13} = \rho[\cos\theta_2(1-\cos\theta_1) + \sin\theta_2\sin\theta_1 + 1 - \cos\theta_2]$$

$$= \rho[1 - \cos(\theta_2+\theta_1)]. \qquad (2.24)$$

Multiplying to obtain the second row yields:

$$N_{21} = -\frac{1}{\rho}(\sin\theta_2\cos\theta_1 + \cos\theta_2\sin\theta_1) = -\frac{1}{\rho}\sin(\theta_2+\theta_1), \qquad (2.25)$$

$$N_{22} = -\sin\theta_2\sin\theta_1 + \cos\theta_2\cos\theta_1 = \cos(\theta_2+\theta_1), \qquad (2.26)$$

$$N_{23} = -\sin\theta_2(1-\cos\theta_1) + \cos\theta_2\sin\theta_1 + \sin\theta_2 = \sin(\theta_2+\theta_1), \quad (2.27)$$

and the last row is obvious, so the horizontal plane works.

2.5   Problem 2–5: RHIC ion parameters

The RHIC collider collides fully stripped gold ions ($A = 197$, $Z = 79$) at a total energy of 100 GeV/nucl. per beam. The circumference of each ring is 3834 m.
(Assume the mass of a gold ion is 197×0.93113 GeV/c^2.)

a) If the injection energy is 10.5 GeV/nucleon, what is the required swing in revolution frequency during acceleration?
b) If we assume that there are 192 identical dipoles per ring, what is the field at top field? Assume each dipole is 10 m long.

Solution:

a)

$$mc^2 = 197 \times 0.93113 \text{ GeV} = 183.43 \text{ GeV} \tag{2.28}$$

$$U_{\text{inj}} = 197 \times 10.5 \text{ GeV} = 2068.5 \text{ GeV} \tag{2.29}$$

$$\gamma_{\text{inj}} = \frac{U_{\text{inj}}}{mc^2} = 11.277 \tag{2.30}$$

$$\beta_{\text{inj}} = \sqrt{1 - \gamma^{-2}} = 0.996060 \tag{2.31}$$

$$f_{\text{inj}} = \frac{0.99606 \times 299792458 \text{ m/s}}{3834 \text{ m}} = 77885 \text{ Hz} \tag{2.32}$$

$$\gamma_{\text{store}} = \frac{197 \times 100 \text{ GeV}}{183.43 \text{ GeV}} = 107.40 \tag{2.33}$$

$$\beta_{\text{store}} = \sqrt{1 - \gamma^{-2}} = 0.999957 \tag{2.34}$$

$$f_{\text{store}} = \frac{0.99996 \times 299792458 \text{ m/s}}{3834 \text{ m}} = 78190 \text{ Hz} \tag{2.35}$$

The frequency swing from injection to top energy is then from 77885 to 78190 Hz (just under 0.4%).
b)

$$p_{\text{store}} = \beta_{\text{store}} \gamma_{\text{store}} \, mc^2 = \sqrt{\gamma_{\text{store}}^2 - 1} \, mc = 19699 \text{ GeV}/c. \tag{2.36}$$

The peak rigidity is

$$\frac{p_{\text{store}}}{q} = \frac{\frac{1.9699 \times 10^{13} \text{ eV}}{2.9979 \times 10^8 \text{ m/s}}}{79e} = 831.76 \text{ T} \cdot \text{m} \tag{2.37}$$

The bend angle per dipole is $\theta = 2\pi/192 = 32.7249$ mr so the bending radius in each dipole must be

$$\rho = \frac{10 \text{ m}}{\theta} = 305.56 \text{ m}. \tag{2.38}$$

Dividing the rigidity by ρ gives peak field in the dipole magnets of

$$B_{\text{store}} \doteq 2.7219 \text{ T}. \tag{2.39}$$

Chapter 3

Problems of Chapter 3: Mechanics of Trajectories

3.1 Problem 3–1: Canonical transformation to local system

The transformation of coordinates from the fixed system to the local system moving along the design orbit is given by the equations (See Fig. 3.1)

$$x = \sqrt{\xi^2 + \eta^2} - \rho, \quad \text{and} \tag{3.1}$$

$$s = \rho \tan^{-1}\left(\frac{\eta}{\xi}\right). \tag{3.2}$$

Show that the momenta given by

$$p_x = p_r = \vec{p} \cdot \hat{x} \quad \text{and} \tag{3.3}$$

$$p_s = \left(1 + \frac{x}{\rho}\right) \vec{p} \cdot \hat{s} \tag{3.4}$$

are canonically conjugate to the coordinates x and s. Hint: Construct a generating function $F_3(p_\xi, p_\eta, x, s)$ as described in Appendix C [of CM].

Solution:

From Fig. 3.1 we have

$$\theta = \frac{s}{\rho}, \tag{3.5}$$

$$\xi = (\rho + x) \cos \theta, \tag{3.6}$$

$$\eta = (\rho + x) \sin \theta, \tag{3.7}$$

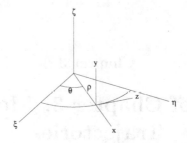

Fig. 3.1   (Fig. 3.2 of CM) Relative to a fixed coordinate system, (ξ, η, y), centered on the center of curvature of the design trajectory, the particle has cylindrical coordinates (r, θ, y). Note that the local system is right-handed if the coordinates are ordered as (x, z, y). The origin of the local system moves with the design particle.

and we want to construct $F_3(\xi, \eta, P_x, P_s)$ (see CM: Appendix C, or Chapter 9 of Ref. [29]) with

$$\xi = (\rho + x) \cos \frac{s}{\rho} = -\frac{\partial F_3}{\partial p_\xi}, \tag{3.8}$$

$$\eta = (\rho + x) \sin \frac{s}{\rho} = -\frac{\partial F_3}{\partial p_\eta}, \tag{3.9}$$

Integrating the two previous partial derivatives and comparing constants yields:

$$F_3(p_\xi, p_\eta, x, s) = -(\rho + x)\left(p_\xi \cos \frac{s}{\rho} + p_\eta \sin \frac{s}{\rho} \right), \tag{3.10}$$

so

$$P_\xi = -\frac{\partial F_3}{\partial x} = p_\xi \cos \theta + p_\eta \sin \theta$$

$$= p_r = \vec{p} \cdot \hat{x}, \tag{3.11}$$

$$P_s = -\frac{\partial F_3}{\partial s} = \left(1 + \frac{x}{\rho}\right)(-p_\xi \sin \theta + p_\eta \cos \theta)$$

$$= \left(1 + \frac{x}{\rho}\right)\vec{p} \cdot \hat{s}. \tag{3.12}$$

3.2 Problem 3–2: Inverse of a symplectic matrix

Prove that the inverse of a symplectic matrix $\mathbf{M}$ is given by

$$\mathbf{M}^{-1} = -\mathbf{S}\mathbf{M}^{\mathrm{T}}\mathbf{S}. \tag{3.13}$$

If we write a symplectic $\mathbf{M}$ in terms of 2×2-blocks $\mathbf{A}$, $\mathbf{B}$, $\mathbf{C}$, and $\mathbf{D}$ as

$$\mathbf{M} = \begin{pmatrix} \mathbf{A}\ \mathbf{B} \\ \mathbf{C}\ \mathbf{D} \end{pmatrix}, \tag{3.14}$$

show that

$$\mathbf{M}^{-1} = \begin{pmatrix} \tilde{\mathbf{A}}\ \tilde{\mathbf{C}} \\ \tilde{\mathbf{B}}\ \tilde{\mathbf{D}} \end{pmatrix}. \tag{3.15}$$

Find a similar formula for a 6×6 symplectic matrix.

Solution:

Starting with the defining equation for a symplectic matrix we have

$$\mathbf{S} = \mathbf{M}^{\mathrm{T}}\mathbf{S}\mathbf{M}. \tag{CM 3.66}$$

Multiplying by $\mathbf{S}^{-1}$ from the left and $\mathbf{M}^{-1}$ from the right gives

$$\mathbf{M}^{-1} = \mathbf{S}^{-1}\mathbf{M}^{\mathrm{T}}\mathbf{S} = -\mathbf{S}\mathbf{M}^{\mathrm{T}}\mathbf{S}. \tag{3.16}$$

Now define the 2×2 block matrices:

$$\mathbf{S}_2 = \begin{pmatrix} 0\ 1 \\ -1\ 0 \end{pmatrix}, \quad \text{and} \quad \mathbf{0} = \begin{pmatrix} 0\ 0 \\ 0\ 0 \end{pmatrix}. \tag{3.17}$$

For the 4×4 case we write $\mathbf{M}$ with 2×2 block submatrices.

$$\begin{aligned} \mathbf{M}^{-1} &= -\begin{pmatrix} \mathbf{S}_2\ \mathbf{0} \\ \mathbf{0}\ \mathbf{S}_2 \end{pmatrix} \begin{pmatrix} \mathbf{A}\ \mathbf{B} \\ \mathbf{C}\ \mathbf{D} \end{pmatrix}^{\mathrm{T}} \begin{pmatrix} \mathbf{S}_2\ \mathbf{0} \\ \mathbf{0}\ \mathbf{S}_2 \end{pmatrix} \\ &= -\begin{pmatrix} \mathbf{S}_2\mathbf{A}^{\mathrm{T}}\mathbf{S}_2\ \mathbf{S}_2\mathbf{C}^{\mathrm{T}}\mathbf{S}_2 \\ \mathbf{S}_2\mathbf{B}^{\mathrm{T}}\mathbf{S}_2\ \mathbf{S}_2\mathbf{D}^{\mathrm{T}}\mathbf{S}_2 \end{pmatrix} = \begin{pmatrix} \tilde{\mathbf{A}}\ \tilde{\mathbf{C}} \\ \tilde{\mathbf{B}}\ \tilde{\mathbf{D}} \end{pmatrix}, \end{aligned} \tag{3.18}$$

having recalled our definition of the symplectic conjugate of a matrix $\mathbf{N}$ from CM: p. 48:

$$\tilde{\mathbf{N}} = \mathbf{S}^{\mathrm{T}}\mathbf{N}^{\mathrm{T}}\mathbf{S} = \mathbf{S}\mathbf{N}^{\mathrm{T}}\mathbf{S}^{\mathrm{T}}. \tag{3.19}$$

For the 6×6 case, we can write the symplectic matrix as

$$\mathbf{M} = \begin{pmatrix} \mathbf{A}\ \mathbf{B}\ \mathbf{C} \\ \mathbf{D}\ \mathbf{E}\ \mathbf{F} \\ \mathbf{G}\ \mathbf{H}\ \mathbf{K} \end{pmatrix}, \tag{3.20}$$

with the nine 2×2 blocks as shown, and we find

$$\mathbf{M}^{-1} = -\mathbf{S}\mathbf{M}^{\mathrm{T}}\mathbf{S} = \tilde{\mathbf{M}} = \begin{pmatrix} \tilde{\mathbf{A}}\ \tilde{\mathbf{D}}\ \tilde{\mathbf{G}} \\ \tilde{\mathbf{B}}\ \tilde{\mathbf{E}}\ \tilde{\mathbf{H}} \\ \tilde{\mathbf{C}}\ \tilde{\mathbf{F}}\ \tilde{\mathbf{K}} \end{pmatrix}. \tag{3.21}$$

3.3   Problem 3–3: Definition of a group

Demonstrate that the collection of real $2n$-dimensional symplectic matrices, $\mathrm{Sp}(2n, \mathbb{R}; \mathbf{S})$, forms a group. The properties of a group[16][28] are:

i for any two elements $a, b \in G$, then $ab \in G$;

ii if a, b, $c \in G$, then $a(bc) = (ab)c$;

iii there is a unique element $e \in G$ such that $ea = a = ae$ for any element $a \in G$;

iv for each $a \in G$ there is an element $a^{-1} \in G$ such that $a^{-1}a = e = aa^{-1}$.

Solution:

i Closure property:

If a and b are symplectic matrices, then using Eq. (CM: 3.66)

$$\mathbf{S} = \mathbf{M}^{\mathrm{T}}\mathbf{S}\mathbf{M}, \tag{CM: 3.66}$$

we must have

$$(ab)^{\mathrm{T}}\mathbf{S}(ab) = b^{\mathrm{T}}(a^{\mathrm{T}}\mathbf{S}a)b = b^{\mathrm{T}}\mathbf{S}b = \mathbf{S} \tag{3.22}$$

since $a^{\mathrm{T}}\mathbf{S}a = \mathbf{S}$ and $b^{\mathrm{T}}\mathbf{S}b = \mathbf{S}$.

ii Associative property:

This follows immediately from the associative nature of matrix multiplication.

iii Existence of a unique identity element:

For symplectic matrices the usual identity element $\mathbf{I}$ must satisfy $a\mathbf{I} = \mathbf{I} = \mathbf{I}a$. Clearly $\mathbf{I} \in G$, since $\mathbf{I}^{\mathrm{T}}\mathbf{S}\mathbf{I} = \mathbf{S}$. To show that the identity element is unique, assume that two different identity elements e_1 and e_2 exist. Multiplying the two gives $e_1 e_2 = e_1 = e_2$ so they must be identical.

iv Existence of inverse elements:

Since the determinant of any symplectic matrix a is nonzero, its inverse a^{-1} must exist. Starting with $a^{\mathrm{T}}\mathbf{S}a = \mathbf{S}$, we multiply by $\mathbf{S}^{\mathrm{T}} = \mathbf{S}^{-1}$ from the left and by a^{-1} to obtain

$$a^{-1} = \mathbf{S}^{\mathrm{T}}a^{\mathrm{T}}\mathbf{S}. \tag{3.23}$$

(Of course we could have just invoked Eq. (3.16) from the previous problem.) Multiplying by a from the left and $\mathbf{S}^{\mathrm{T}}$ from the right yields

$$\mathbf{S}^{\mathrm{T}} = a\mathbf{S}^{\mathrm{T}}a^{\mathrm{T}}. \tag{3.24}$$

Taking the transpose of both sides, we find that a^{T} is also symplectic:

$$\mathbf{S} = a\mathbf{S}a^{\mathrm{T}}. \tag{3.25}$$

We must show that a^{-1} is also symplectic:

$$(a^{-1})^{\mathrm{T}}\mathbf{S}a^{-1} = (\mathbf{S}^{\mathrm{T}}a^{\mathrm{T}}\mathbf{S})^{\mathrm{T}}\mathbf{S}(\mathbf{S}^{\mathrm{T}}a^{\mathrm{T}}\mathbf{S}) = \mathbf{S}^{\mathrm{T}}a\mathbf{S}(\mathbf{S}\mathbf{S}^{\mathrm{T}})a^{\mathrm{T}}\mathbf{S}$$
$$= \mathbf{S}^{\mathrm{T}}(a\mathbf{S}a^{\mathrm{T}})\mathbf{S} = (\mathbf{S}^{\mathrm{T}}\mathbf{S})\mathbf{S} = \mathbf{S}. \tag{3.26}$$

3.4   Problem 3–4:   Unitary similarity transformations of Sp($2n$)

The usual representation for the Pauli spin matrices of SU(2) is:

$$\sigma_x = \begin{pmatrix} 0 & 1 \\ 1 & 0 \end{pmatrix}, \quad \sigma_y = \begin{pmatrix} 0 & -i \\ i & 0 \end{pmatrix}, \quad \sigma_z = \begin{pmatrix} 1 & 0 \\ 0 & -1 \end{pmatrix}. \qquad (3.27)$$

For one degree of freedom, $\mathbf{S} = i\sigma_y$. This suggests a connection between SU(2) and Sp($2n$).

a) Given a matrix $\mathbf{M}$ for some beam line, show that the transformation defined by

$$\check{\mathbf{M}} = \sigma_z \mathbf{M}^{-1} \sigma_z, \qquad (3.28)$$

is the transfer matrix for the mirror image of the beam line. (For example, if

$$\mathbf{M} = \mathbf{M}_q \mathbf{M}_b \mathbf{M}_d, \qquad (3.29)$$

for a drift followed by a bend and then a quadrupole, then

$$\check{\mathbf{M}} = \mathbf{M}_d \mathbf{M}_b \mathbf{M}_q.) \qquad (3.30)$$

Note that if the independent variable used is time, then σ_z is the time reversal operator.

b) Show that the following relations hold:

$$\mathbf{M}^{\mathrm{T}} = \sigma_x \check{\mathbf{M}} \sigma_x, \qquad (3.31)$$

$$\mathbf{M}^{-1} = \sigma_y \mathbf{M}^{\mathrm{T}} \sigma_y. \qquad (3.32)$$

Solution:

a) Reflection operation:

The inverse operation on $\mathbf{M}$ reverses the direction of particle propagation from the last element to the first; however time is also reversed, so that a particle is returned to its initial point in phase space, since $\mathbf{M}^{-1}\mathbf{M} = \mathbf{I}$. A time reversal operator Θ operating on a point in phase space reverses the direction of motion ($\Theta\vec{P} = -\vec{P}$) but leaves the spatial coordinates unchanged ($\Theta\vec{X} = \vec{X}$). If we start with a Hamiltonian $H(x, P_x, y, P_y, s, P_s; t)$ with time as the independent parameter, then it is easy to see that σ_z is a time reversal operator for each canonical pair, since

$$\begin{pmatrix} x_j \\ -P_j \end{pmatrix} = \begin{pmatrix} 1 & 0 \\ 0 & -1 \end{pmatrix} \begin{pmatrix} x_j \\ P_j \end{pmatrix}. \qquad (3.33)$$

For the full 6×6 matrix with t as the independent parameter, we may define the time reversal operator in 2×2-block form as

$$\Theta = \begin{pmatrix} \sigma_z & \mathbf{0} & \mathbf{0} \\ \mathbf{0} & \sigma_z & \mathbf{0} \\ \mathbf{0} & \mathbf{0} & \sigma_z \end{pmatrix}. \tag{3.34}$$

We should also note that $\Theta^2 = \mathbf{I}$ as should expected for a time reversal operator, so $\Theta^{-1} = \Theta$. Applying a similarity transformation of the time reversal operator to $\mathbf{M}^{-1}$, should give the desired result $\check{\mathbf{M}} = \Theta\mathbf{M}^{-1}\Theta$. If instead we interchange time and the longitudinal coordinate in the Hamiltonian $H(x, P_x, y, P_y, t, -U; s)$, the required time reversal operator becomes

$$\Theta = \begin{pmatrix} \sigma_z & \mathbf{0} & \mathbf{0} \\ \mathbf{0} & \sigma_z & \mathbf{0} \\ \mathbf{0} & \mathbf{0} & -\sigma_z \end{pmatrix}, \tag{3.35}$$

with $\Theta^2 = \mathbf{I}$ as before.

We should note that the diagonal 2×2 blocks of the quadrupole, drift, and bend matrices all have the form (with equal elements on the diagonal):

$$\mathbf{E} = \begin{pmatrix} a & b \\ c & a \end{pmatrix}, \quad \text{with} \quad a^2 - b = 1, \tag{3.36}$$

and since

$$\begin{aligned}
\check{\mathbf{E}} = \sigma_z \mathbf{E}^{-1} \sigma_z &= \begin{pmatrix} 1 & 0 \\ 0 & -1 \end{pmatrix} \begin{pmatrix} a & b \\ c & a \end{pmatrix}^{-1} \begin{pmatrix} 1 & 0 \\ 0 & -1 \end{pmatrix} \\
&\doteq \begin{pmatrix} 1 & 0 \\ 0 & -1 \end{pmatrix} \begin{pmatrix} a & -b \\ -c & a \end{pmatrix} \begin{pmatrix} 1 & 0 \\ 0 & -1 \end{pmatrix} = \begin{pmatrix} a & b \\ c & a \end{pmatrix} = \mathbf{E},
\end{aligned} \tag{3.37}$$

the upper-left 4×4 block of matrices for these simple elements remain unchanged under reflection. Invoking the result of Eq. (3.21), we show that the dispersion terms in the off-diagonal 2×2 blocks of a sector bend also remain unchanged:

$$\begin{aligned}
\sigma_z \mathbf{S} \begin{pmatrix} 0 & \rho(1 - \cos\theta) \\ 0 & \sin\theta \end{pmatrix}^{\mathrm{T}} \mathbf{S}^{\mathrm{T}}(-\sigma_z) &= \begin{pmatrix} 1 & 0 \\ 0 & -1 \end{pmatrix} \begin{pmatrix} 0 & 1 \\ -1 & 0 \end{pmatrix} \begin{pmatrix} 0 & \rho(1 - \cos\theta) \\ 0 & \sin\theta \end{pmatrix}^{\mathrm{T}} \\
&\times \begin{pmatrix} 0 & -1 \\ 1 & 0 \end{pmatrix} \begin{pmatrix} -1 & 0 \\ 0 & 1 \end{pmatrix}
\end{aligned}$$

$$= \begin{pmatrix} 0 & 1 \\ 1 & 0 \end{pmatrix} \begin{pmatrix} 0 & 0 \\ \rho(1-\cos\theta)\sin\theta \end{pmatrix} \begin{pmatrix} 0 & -1 \\ -1 & 0 \end{pmatrix}$$

$$= -\begin{pmatrix} \rho(1-\cos\theta)\sin\theta & 0 \\ 0 & 0 \end{pmatrix} \begin{pmatrix} 0 & 1 \\ 1 & 0 \end{pmatrix}$$

$$= \begin{pmatrix} -\sin\theta & -\rho(1-\cos\theta) \\ 0 & 0 \end{pmatrix}, \qquad (3.38)$$

since $\sigma_z \mathbf{S} = \sigma_z(i\sigma_y) = i(\sigma_z\sigma_y) = i(-i\sigma_x) = \sigma_x$.

A rotation about the z-axis transforms as

$$\check{\mathbf{R}} = \Theta\mathbf{R}^{-1}\Theta = \begin{pmatrix} \sigma_z & 0 & 0 \\ 0 & \sigma_z & 0 \\ 0 & 0 & \pm\sigma_z \end{pmatrix} \begin{pmatrix} \mathbf{I}\cos\theta & \mathbf{I}\sin\theta & 0 \\ -\mathbf{I}\sin\theta & \mathbf{I}\cos\theta & 0 \\ 0 & 0 & \mathbf{I} \end{pmatrix}^{-1} \begin{pmatrix} \sigma_z & 0 & 0 \\ 0 & \sigma_z & 0 \\ 0 & 0 & \pm\sigma_z \end{pmatrix}$$

$$= \begin{pmatrix} \mathbf{I}\cos\theta & -\mathbf{I}\sin\theta & 0 \\ \mathbf{I}\sin\theta & \mathbf{I}\cos\theta & 0 \\ 0 & 0 & \mathbf{I} \end{pmatrix} = \mathbf{R}^{-1}, \qquad (3.39)$$

so that we find the reflection of a rotated element $\mathbf{E}_r = \mathbf{R}\mathbf{E}\mathbf{R}^{-1}$ is

$$\check{\mathbf{E}}_r = \Theta\mathbf{E}_r^{-1}\Theta = \Theta(\mathbf{R}\mathbf{E}\mathbf{R}^{-1})^{-1}\Theta$$

$$= \Theta\mathbf{R}\mathbf{E}^{-1}\mathbf{R}^{-1}\Theta = (\Theta\mathbf{R}\Theta)(\Theta\mathbf{E}^{-1}\Theta)(\Theta\mathbf{R}^{-1}\Theta)$$

$$= \mathbf{R}^{-1}\check{\mathbf{E}}\mathbf{R} \qquad (3.40)$$

with the mirrored element rotated in the opposite direction as should be exected.

b) Connections of reflection to transpose and inverse operations:

Recalling that the Pauli matrices satisfy the cyclic relations $\sigma_j\sigma_k = i\epsilon_{jkl}\sigma_l$, we see that

$$\sigma_x\check{\mathbf{M}}\sigma_x = \sigma_x\left(\sigma_z\mathbf{M}^{-1}\sigma_z\right)\sigma_x = (-i\sigma_y)\,\mathbf{M}^{-1}(i\sigma_y). \qquad (3.41)$$

Since

$$i\sigma_y = \begin{pmatrix} 0 & 1 \\ -1 & 0 \end{pmatrix} = \mathbf{S}_2, \qquad (3.42)$$

we have

$$\mathbf{M}^{-1} = \mathbf{S}_2^{\mathrm{T}}\mathbf{M}^{\mathrm{T}}\mathbf{S}_2 = (-i\sigma_y)\mathbf{M}^{\mathrm{T}}(i\sigma_y) = \sigma_y\mathbf{M}^{\mathrm{T}}\sigma_y, \qquad (3.43)$$

for the relation between the inverse and transpose matrices, and

$$\sigma_x\check{\mathbf{M}}\sigma_x = -\mathbf{S}_2\mathbf{M}^{-1}\mathbf{S}_2 = -\mathbf{S}_2\left(\mathbf{S}_2^{\mathrm{T}}\mathbf{M}^{\mathrm{T}}\mathbf{S}_2\right)\mathbf{S}_2 = \mathbf{M}^{\mathrm{T}}, \qquad (3.44)$$

for the transformation for a reflected beam-line matrix to the transpose of the matrix.

3.5   Problem 3–5: Drift matrix

Use the symplectic generator method to find the transfer matrix for a drift of length L.

Solution:

For a drift element with no magnetic field, the Hamiltonian of Eq. (CM:3.90) becomes

$$\mathcal{H} = -\left(1 + \delta - \frac{1}{2}(x'^2 + y'^2) + \cdots\right) + \delta + 1 = \frac{1}{2}(x'^2 + y'^2) + \cdots . \quad (3.45)$$

The equations of motion are then

$$\frac{dx}{ds} = \frac{\partial \mathcal{H}}{\partial x'} \simeq x', \qquad \frac{dx'}{ds} = -\frac{\partial \mathcal{H}}{\partial x} \simeq 0, \qquad (3.46)$$

$$\frac{dy}{ds} = \frac{\partial \mathcal{H}}{\partial y'} \simeq y', \qquad \frac{dy'}{ds} = -\frac{\partial \mathcal{H}}{\partial y} \simeq 0, \qquad (3.47)$$

$$\frac{dz}{ds} = \frac{\partial \mathcal{H}}{\partial \delta} \simeq 0, \qquad \frac{d\delta}{ds} = -\frac{\partial \mathcal{H}}{\partial z} \simeq 0, \qquad (3.48)$$

and the linearized vector difference increments are

$$\mathbf{K}ds = \begin{pmatrix} 0 & 1 & 0 & 0 & 0 & 0 \\ 0 & 0 & 0 & 0 & 0 & 0 \\ 0 & 0 & 0 & 1 & 0 & 0 \\ 0 & 0 & 0 & 0 & 0 & 0 \\ 0 & 0 & 0 & 0 & 0 & 0 \\ 0 & 0 & 0 & 0 & 0 & 0 \end{pmatrix} ds. \qquad (3.49)$$

Integration via Exponentiation yields

$$\mathbf{M} = \lim_{n \to \infty} \left(\mathbf{I} + \mathbf{K}\frac{L}{n}\right)^n = \mathbf{I} + \mathbf{K}L = \begin{pmatrix} 1 & L & 0 & 0 & 0 & 0 \\ 0 & 1 & 0 & 0 & 0 & 0 \\ 0 & 0 & 1 & L & 0 & 0 \\ 0 & 0 & 0 & 1 & 0 & 0 \\ 0 & 0 & 0 & 0 & 1 & 0 \\ 0 & 0 & 0 & 0 & 0 & 1 \end{pmatrix}, \qquad (3.50)$$

since $\mathbf{K}^2 = \mathbf{0}$.

3.6   Problem 3–6: Symplectic constraints of a bend matrix

For general static magnetic elements with no xy-coupling and no rf electric fields the transport matrix for (x, x', z, δ) has the general form:

$$\begin{pmatrix} C & S & 0 & D \\ C' & S' & 0 & D' \\ E & F & 1 & G \\ 0 & 0 & 0 & 1 \end{pmatrix} \qquad (3.51)$$

a) Using the symplectic condition $\mathbf{M}^{\mathrm{T}}\mathbf{S}\mathbf{M} = \mathbf{S}$ express E and F in terms of the other elements and show that there is no constraint on the value of G.

b) Show that matrices of this form make a subgroup of $\mathrm{Sp}(4)$ the real symplectic group of 4×4 matrices.

Solution:

a)

$$
\mathbf{M}^{\mathrm{T}}\mathbf{S}\mathbf{M} = \begin{pmatrix} C & C' & E & 0 \\ S & S' & F & 0 \\ 0 & 0 & 1 & 0 \\ D & D' & G & 1 \end{pmatrix} \begin{pmatrix} 0 & 1 & 0 & 0 \\ -1 & 0 & 0 & 0 \\ 0 & 0 & 0 & 1 \\ 0 & 0 & -1 & 0 \end{pmatrix} \begin{pmatrix} C & S & 0 & D \\ C' & S' & 0 & D' \\ E & F & 1 & G \\ 0 & 0 & 0 & 1 \end{pmatrix}
$$

$$
= \begin{pmatrix} C & C' & E & 0 \\ S & S' & F & 0 \\ 0 & 0 & 1 & 0 \\ D & D' & G & 1 \end{pmatrix} \begin{pmatrix} C' & S' & 0 & D' \\ -C & -S & 0 & -D \\ 0 & 0 & 0 & 1 \\ -E & -F & -1 & -G \end{pmatrix}
$$

$$
= \begin{pmatrix} 0 & CS' - C'S & 0 & CD' - C'D + E \\ SC' - S'C & 0 & 0 & SD' - S'D + F \\ 0 & 0 & 0 & 1 \\ DC' - D'C - E & DS' - D'S - F & -1 & 0 \end{pmatrix}
$$

$$
= \begin{pmatrix} 0 & 1 & 0 & 0 \\ -1 & 0 & 0 & 0 \\ 0 & 0 & 0 & 1 \\ 0 & 0 & -1 & 0 \end{pmatrix}. \qquad (3.52)
$$

We find the usual symplectic requirement for the upper-left 2×2-block: $CS' - SC' = 1$. The 14 and 41 give the same result of zero for a symmetric matrix provided that we have

$$E = DC' - D'C. \tag{3.53}$$

Similarly the 24 and 42 elements require that

$$F = DS' - D'S. \tag{3.54}$$

Note that there is no G in the third line of Eq. (3.52), so the symplectic property of $\mathbf{M}$ does not put any constraint on the value of $M_{3,4} = G$.

b) The matrices of this subgroup must satisfy the four properties of a group as defined in Problem 3–3. The identity element is clearly of this form with $C = S' = 0$ and $D = D' = E = F = G = 0$. What we need to show is that symplectic matrices of the form Eq. (3.51) satisfy the the closure rule (i. e. that any product of two such matrices has the same form), and that their inverses are also of the same form. Since these bend elements, as being derivable from a Hamiltonian, must be in $\mathrm{Sp}(4)$, the associativity requirement will be inherited from that group provided that the closure property is satisfied.

So for closure we have

$$\mathbf{MN} = \begin{pmatrix} C & S & 0 & D \\ C' & S' & 0 & D' \\ E & F & 1 & G \\ 0 & 0 & 0 & 1 \end{pmatrix} \begin{pmatrix} c & s & 0 & d \\ c' & s' & 0 & d' \\ e & f & 1 & g \\ 0 & 0 & 0 & 1 \end{pmatrix}$$

$$= \begin{pmatrix} Cc + Sc' & Cs + Ss' & 0 & Cd + Sd' + D \\ C'c + S'c' & C's + S's' & 0 & C'dS'd' + D' \\ Ec + Fc' + e & Es + Fs' + f & 1 & Ed + Fd' + G + g \\ 0 & 0 & 0 & 1 \end{pmatrix}, \tag{3.55}$$

which is of the desired form. Note: since $\mathbf{M}$ and $\mathbf{N}$ are both in the symplectic group $\mathrm{Sp}(4)$, then we already know that their product must also be in $\mathrm{Sp}(4)$.

Taking $\mathbf{M}$ given by the matrix in Eq. (3.51), and from Problem 3–2, we have the inverse given by

$$\mathbf{M}^{-1} = -\mathbf{S}\mathbf{M}^\mathrm{T}\mathbf{S}$$

$$= \begin{pmatrix} 0 & -1 & 0 & 0 \\ 1 & 0 & 0 & 0 \\ 0 & 0 & 0 & -1 \\ 0 & 0 & 1 & 0 \end{pmatrix} \begin{pmatrix} C & C' & E & 0 \\ S & S' & F & 0 \\ 0 & 0 & 1 & 0 \\ D & D' & G & 1 \end{pmatrix} \begin{pmatrix} 0 & 1 & 0 & 0 \\ -1 & 0 & 0 & 0 \\ 0 & 0 & 0 & 1 \\ 0 & 0 & -1 & 0 \end{pmatrix}$$

$$= \begin{pmatrix} -S & -S' & -F & 0 \\ C & C' & E & 0 \\ -D & -D' & -G & -1 \\ 0 & 0 & 1 & 0 \end{pmatrix} \begin{pmatrix} 0 & 1 & 0 & 0 \\ -1 & 0 & 0 & 0 \\ 0 & 0 & 0 & 1 \\ 0 & 0 & -1 & 0 \end{pmatrix}$$

$$= \begin{pmatrix} S' & -S & 0 & -F \\ -C' & C & 0 & E \\ D' & -D & 1 & -G \\ 0 & 0 & 0 & 1 \end{pmatrix}, \tag{3.56}$$

which has the desired form.

3.7 Problem 3-7: Quadrupole matrix

Use the symplectic generator method to find the transfer matrix for a quadrupole magnet of length L with $A_x = A_y = 0$ and

$$A_s = \frac{g}{2}(x^2 - y^2). \tag{3.57}$$

Solution:

We can start with the Hamiltonian from Eq. (CM: 3.83) but taking $\rho \to \infty$ for zero field on the central trajectory:

$$\mathcal{H} = -\frac{q}{p_0} A_s + \frac{1}{2}(x'^2 + y'^2) + \cdots$$

$$= -\frac{qg}{2p_0}(x^2 - y^2) + \frac{1}{2}(x'^2 + y'^2) + \cdots . \tag{3.58}$$

$$\tag{3.59}$$

Hamilton's equations are then

$$\frac{dx}{ds} = \frac{\partial \mathcal{H}}{\partial x'} \simeq x', \tag{3.60}$$

$$\frac{dy}{ds} = \frac{\partial \mathcal{H}}{\partial y'} \simeq y', \tag{3.61}$$

$$\frac{dz}{ds} = \frac{\partial \mathcal{H}}{\partial \delta} = 0,^1 \tag{3.62}$$

$$\frac{dx'}{ds} = -\frac{\partial \mathcal{H}}{\partial x} \simeq -\frac{qg}{p_0} x, \tag{3.63}$$

$$\frac{dy'}{ds} = -\frac{\partial \mathcal{H}}{\partial y} \simeq +\frac{qg}{p_0} y, \tag{3.64}$$

$$\frac{d\delta}{ds} = 0. \tag{3.65}$$

Now we construct the finite difference equation:

$$\begin{pmatrix} x_1 \\ x_1' \\ y_1 \\ y_1' \\ z_1 \\ \delta_1 \end{pmatrix} = \begin{pmatrix} 1 & ds & 0 & 0 & 0 & 0 \\ -\frac{qg}{p_0} ds & 1 & 0 & 0 & 0 & 0 \\ 0 & 0 & 1 & ds & 0 & 0 \\ 0 & 0 & \frac{qg}{p_0} ds & 1 & 0 & 0 \\ 0 & 0 & 0 & 0 & 1 & 0 \\ 0 & 0 & 0 & 0 & 0 & 1 \end{pmatrix} \begin{pmatrix} x_0 \\ x_0' \\ y_0 \\ y_0' \\ z_0 \\ \delta_0 \end{pmatrix} = (\mathbf{I} + \mathbf{G}\, ds) \begin{pmatrix} x_0 \\ x_0' \\ y_0 \\ y_0' \\ z_0 \\ \delta_0 \end{pmatrix}. \tag{3.66}$$

---

[1]This is actually only true in the ultrarelativistic limit. See the discussion following Eq. (3.102) and § 7.4 for a more correct treatment of the M_{56} element.

Define $k = \frac{qg}{p_0}$, and

$$\mathbf{K} = \mathbf{G} = \sqrt{k} \begin{pmatrix} 0 & \frac{1}{\sqrt{k}} & 0 & 0 & 0 & 0 \\ -\sqrt{k} & 0 & 0 & 0 & 0 & 0 \\ 0 & 0 & 0 & \frac{1}{\sqrt{k}} & 0 & 0 \\ 0 & 0 & \sqrt{k} & 0 & 0 & 0 \\ 0 & 0 & 0 & 0 & 0 & 0 \\ 0 & 0 & 0 & 0 & 0 & 0 \end{pmatrix}. \tag{3.67}$$

Raising $\mathbf{K}$ to successive powers yields

$$\mathbf{K}^2 = k \begin{pmatrix} -1 & 0 & 0 & 0 & 0 & 0 \\ 0 & -1 & 0 & 0 & 0 & 0 \\ 0 & 0 & 1 & 0 & 0 & 0 \\ 0 & 0 & 0 & 1 & 0 & 0 \\ 0 & 0 & 0 & 0 & 0 & 0 \\ 0 & 0 & 0 & 0 & 0 & 0 \end{pmatrix}, \tag{3.68}$$

$$\mathbf{K}^3 = k^{3/2} \begin{pmatrix} 0 & -\frac{1}{\sqrt{k}} & 0 & 0 & 0 & 0 \\ \sqrt{k} & 0 & 0 & 0 & 0 & 0 \\ 0 & 0 & 0 & \frac{1}{\sqrt{k}} & 0 & 0 \\ 0 & 0 & \sqrt{k} & 0 & 0 & 0 \\ 0 & 0 & 0 & 0 & 0 & 0 \\ 0 & 0 & 0 & 0 & 0 & 0 \end{pmatrix}, \tag{3.69}$$

and

$$\mathbf{K}^4 = k^2 \begin{pmatrix} 1 & 0 & 0 & 0 & 0 & 0 \\ 0 & 1 & 0 & 0 & 0 & 0 \\ 0 & 0 & 1 & 0 & 0 & 0 \\ 0 & 0 & 0 & 1 & 0 & 0 \\ 0 & 0 & 0 & 0 & 0 & 0 \\ 0 & 0 & 0 & 0 & 0 & 0 \end{pmatrix}. \tag{3.70}$$

From this we see the repetitive pattern. The three planes (horizontal xx', vertical yy' and $z\delta$) are decoupled, so we have just nonzero values in the 2×2 blocks along the diagonals.

$$\mathbf{M} = \lim_{n \to \infty} \left(\mathbf{I} + \mathbf{K}\frac{L}{n} \right)^n = e^{\mathbf{K}L}$$

$$= \mathbf{I} + \mathbf{K}L + \frac{\mathbf{K}^2 L^2}{2!} + \frac{\mathbf{K}^3 L^3}{3!} + \cdots \tag{3.71}$$

The xx' diagonal block is

$$\begin{pmatrix} M_{11} & M_{12} \\ M_{21} & M_{22} \end{pmatrix} = \begin{pmatrix} 1 & 0 \\ 0 & 1 \end{pmatrix} + \sqrt{k}L \begin{pmatrix} 0 & \frac{1}{\sqrt{k}} \\ -\sqrt{k} & 0 \end{pmatrix} - \frac{\left(\sqrt{k}L\right)^2}{2!} \begin{pmatrix} 1 & 0 \\ 0 & 1 \end{pmatrix}$$

$$- \frac{\left(\sqrt{k}L\right)^3}{3!} \begin{pmatrix} 0 & \frac{1}{\sqrt{k}} \\ -\sqrt{k} & 0 \end{pmatrix} + \frac{\left(\sqrt{k}L\right)^4}{4!} \begin{pmatrix} 1 & 0 \\ 0 & 1 \end{pmatrix} + \cdots, \tag{3.72}$$

or upon collecting terms the components become

$$M_{11} = M_{22} = 1 - \frac{\left(\sqrt{k}\,L\right)^2}{2!} + \frac{\left(\sqrt{k}\,L\right)^4}{4!} - \cdots = \cos\left(\sqrt{k}\,L\right), \tag{3.73}$$

$$M_{12} = \frac{1}{\sqrt{k}}\left[\sqrt{k}L - \frac{1}{3!}\left(\sqrt{k}L\right)^3 + \cdots\right] = \frac{1}{\sqrt{k}}\sin\left(\sqrt{k}L\right), \tag{3.74}$$

$$M_{21} = -\sqrt{k}\left[\sqrt{k}L - \frac{1}{3!}\left(\sqrt{k}L\right)^3 + \cdots\right] = -\sqrt{k}\sin\left(\sqrt{k}L\right). \tag{3.75}$$

Similarly, the yy' diagonal block is

$$\begin{pmatrix} M_{33} & M_{34} \\ M_{43} & M_{44} \end{pmatrix} = \begin{pmatrix} 1 & 0 \\ 0 & 1 \end{pmatrix} + \sqrt{k}L \begin{pmatrix} 0 & \frac{1}{\sqrt{k}} \\ -\sqrt{k} & 0 \end{pmatrix} + \frac{\left(\sqrt{k}L\right)^2}{2!} \begin{pmatrix} 1 & 0 \\ 0 & 1 \end{pmatrix}$$

$$+ \frac{\left(\sqrt{k}L\right)^3}{3!} \begin{pmatrix} 0 & \frac{1}{\sqrt{k}} \\ -\sqrt{k} & 0 \end{pmatrix} + \frac{\left(\sqrt{k}L\right)^4}{4!} \begin{pmatrix} 1 & 0 \\ 0 & 1 \end{pmatrix} + \cdots, \tag{3.76}$$

which yields hyperbolic functions rather than the circular functions:

$$M_{33} = M_{44} = 1 + \frac{\left(\sqrt{k}\,L\right)^2}{2!} + \frac{\left(\sqrt{k}\,L\right)^4}{4!} + \cdots = \cosh\left(\sqrt{k}\,L\right), \tag{3.77}$$

$$M_{34} = \frac{1}{\sqrt{k}}\left[\sqrt{k}L + \frac{1}{3!}\left(\sqrt{k}L\right)^3 + \cdots\right] = \frac{1}{\sqrt{k}}\sinh\left(\sqrt{k}L\right), \tag{3.78}$$

$$M_{43} = \sqrt{k}\left[\sqrt{k}L + \frac{1}{3!}\left(\sqrt{k}L\right)^3 + \cdots\right] = \sqrt{k}\sinh\left(\sqrt{k}L\right), \tag{3.79}$$

The $z\delta$ diagonal block is just the 2×2 identity matrix. Putting all this together gives the 6×6 matrix

$$\begin{pmatrix} \cos\left(\sqrt{k}L\right) & \frac{1}{\sqrt{k}}\sin\left(\sqrt{k}L\right) & 0 & 0 & 0 & 0 \\ -\sqrt{k}\sin\left(\sqrt{k}L\right) & \cos\left(\sqrt{k}L\right) & 0 & 0 & 0 & 0 \\ 0 & 0 & \cosh\left(\sqrt{k}L\right) & \frac{1}{\sqrt{k}}\sinh\left(\sqrt{k}L\right) & 0 & 0 \\ 0 & 0 & \sqrt{k}\sinh\left(\sqrt{k}L\right) & \cosh\left(\sqrt{k}L\right) & 0 & 0 \\ 0 & 0 & 0 & 0 & 1 & 0 \\ 0 & 0 & 0 & 0 & 0 & 1 \end{pmatrix}.$$

$$\tag{3.80}$$

3.8   Problem 3–8: Solenoid matrix

a) Find the equations of motion for a particle in a uniform magnetic field along the s-axis, with the vector potential

$$\vec{A} = \begin{cases} \left(-\frac{B_0}{2}y, \frac{B_0}{2}x, 0\right), & \text{if } 0 < s < l; \\ 0, & \text{otherwise.} \end{cases} \tag{3.81}$$

Be sure to use a canonical system of coordinates inside the magnet.

b) Find the generator **G** and use it to obtain the linear transformation matrix,

$$\mathbf{M} = \begin{pmatrix} \frac{1+\cos\phi}{2} & r\sin\phi & \frac{\sin\phi}{2} & r(1-\cos\phi) \\ -\frac{\sin\phi}{4r} & \frac{1+\cos\phi}{2} & -\frac{1-\cos\phi}{4r} & \frac{\sin\phi}{2} \\ -\frac{\sin\phi}{2} & -r(1-\cos\phi) & \frac{1+\cos\phi}{2} & r\sin\phi \\ \frac{1-\cos\phi}{4r} & -\frac{\sin\phi}{2} & -\frac{\sin\phi}{4r} & \frac{1+\cos\phi}{2} \end{pmatrix}, \tag{3.82}$$

for the transformation through a solenoid magnet in the hard-edge approximation, with $r = p_s/(qB_0)$, and $\phi = l/r$.

Solution:

a) Since the vector potential has transverse components rather than just a longitudinal component, we must start with Eq. (CM: 3.70) for the Hamiltonian:

$$\frac{H}{p_0} = -\frac{qA_s}{p_0} - \left(1 + \frac{x}{\rho}\right)$$
$$\sqrt{\left(\frac{U}{p_0c}\right)^2 - \left(\frac{mc}{p_0}\right)^2 - \left(\frac{P_x - qA_x}{p_0}\right)^2 - \left(\frac{P_y - qA_y}{p_0}\right)^2}. \tag{CM:3.70}$$

In order to simplify the calculations a little, let us rescale the canonical momenta $w_x = P_x/p_0$, $w_y = P_y/p_0$, and $w_t = U/(p_0c)$, since the design momentum p_0 is just a constant. Note that outside the solenoid ($s < 0$ or $s > l$) and $w_x = p_x/p_0 \simeq x'$ and $w_y = p_y/p_0 \simeq y'$ in the paraxial limit. The bending radius for the design particle travelling down the axis of the solenoid is infinite, so our new Hamiltonian becomes

$$\mathcal{H}(x, w_x, y, w_y, t, -w_t; s) = \frac{H}{p_0} = -\frac{p_s}{p_0}$$
$$= -\sqrt{w_t^2 - \left(\frac{mc}{p_0}\right)^2 - \left(w_x - \frac{qA_x}{p_0}\right)^2 - \left(w_y - \frac{qA_y}{p_0}\right)^2}. \tag{3.83}$$

Hamilton's equations of motion are then

$$\frac{dx}{ds} = \frac{d\mathcal{H}}{dw_x} = \frac{w_x - \frac{qA_x}{p_0}}{\sqrt{w_t^2 - \left(\frac{mc}{p_0}\right)^2 - \left(w_x - \frac{qA_x}{p_0}\right)^2 - \left(w_y - \frac{qA_y}{p_0}\right)^2}}, \quad (3.84)$$

$$\frac{dy}{ds} = \frac{d\mathcal{H}}{dw_y} = \frac{w_y - \frac{qA_y}{p_0}}{\sqrt{w_t^2 - \left(\frac{mc}{p_0}\right)^2 - \left(w_x - \frac{qA_x}{p_0}\right)^2 - \left(w_y - \frac{qA_y}{p_0}\right)^2}}, \quad (3.85)$$

$$\frac{dw_x}{ds} = -\frac{d\mathcal{H}}{dx} = \frac{\left(w_x - \frac{qA_x}{p_0}\right)\frac{q}{p_0}\frac{\partial A_x}{\partial x} + \left(w_y - \frac{qA_y}{p_0}\right)\frac{q}{p_0}\frac{\partial A_y}{\partial x}}{\sqrt{w_t^2 - \left(\frac{mc}{p_0}\right)^2 - \left(w_x - \frac{qA_x}{p_0}\right)^2 - \left(w_y - \frac{qA_y}{p_0}\right)^2}},$$
$$(3.86)$$

$$\frac{dw_y}{ds} = -\frac{d\mathcal{H}}{dy} = \frac{\left(w_x - \frac{qA_x}{p_0}\right)\frac{q}{p_0}\frac{\partial A_x}{\partial y} + \left(w_y - \frac{qA_y}{p_0}\right)\frac{q}{p_0}\frac{\partial A_y}{\partial y}}{\sqrt{w_t^2 - \left(\frac{mc}{p_0}\right)^2 - \left(w_x - \frac{qA_x}{p_0}\right)^2 - \left(w_y - \frac{qA_y}{p_0}\right)^2}},$$
$$(3.87)$$

for the transverse equations of motion. Incidentally, the longitudinal equations are

$$\frac{dt}{ds} = -\frac{d\mathcal{H}}{dw_t} = \frac{-w_t}{\sqrt{w_t^2 - \left(\frac{mc}{p_0}\right)^2 - \left(w_x - \frac{qA_x}{p_0}\right)^2 - \left(w_y - \frac{qA_y}{p_0}\right)^2}},$$
$$(3.88)$$

$$\frac{dw_t}{ds} = \frac{d\mathcal{H}}{dt} = 0. \quad (3.89)$$

The partial derivatives of the vector potential components are

$$\frac{\partial A_x}{\partial x} = 0, \quad (3.90)$$

$$\frac{\partial A_x}{\partial y} = -B_0/2, \quad (3.91)$$

$$\frac{\partial A_y}{\partial x} = B_0/2, \quad (3.92)$$

$$\frac{\partial A_y}{\partial 0} = 0. \quad (3.93)$$

What we want is to find the equations of motion to first order in the canonical variables. We need to expand $w_t = w_{t0} + \Delta w_t$, where

$$w_{t0} = \frac{U}{p_0 c} = \frac{\gamma_0 mc^2}{\gamma_0 \beta_0 mc^2} = \frac{1}{\beta_0}, \quad (3.94)$$

and

$$w_{t0}^2 - \frac{m^2c^2}{p_0^2} = \frac{1}{\beta_0^2} - \frac{1}{\gamma_0^2\beta_0^2} = 1. \tag{3.95}$$

So dw_t/ds becomes just $d(\Delta w_t)/ds$, and

$$w_t^2 - \left(\frac{mc}{p_0}\right)^2 = (w_{t0} + \Delta w_t)^2 - \left(\frac{mc}{p_0}\right)^2 = 1 + \frac{2\Delta w_t}{\beta_0} + (\Delta w_t)^2. \tag{3.96}$$

Keeping only the first order terms, the reciprocal of the square root to first order in Δw_t becomes

$$\frac{1}{(\sqrt{\cdots})} = 1 - \frac{\Delta w_t}{\beta_0} + \mathcal{O}(2). \tag{3.97}$$

The four transverse equations of motion to first order are

$$\frac{dx}{ds} = \frac{p_0}{p_s}\left(w_x + \frac{qB_0}{2p_0}y\right), \tag{3.98}$$

$$\frac{dw_x}{ds} = \frac{p_0}{p_s}\frac{qB_0}{2p_0}\left(w_y - \frac{qB_0}{2p_0}x\right) + \mathcal{O}(2), \tag{3.99}$$

$$\frac{dy}{ds} = \frac{p_0}{p_s}\left(w_y - \frac{qB_0}{2p_0}x\right), \tag{3.100}$$

$$\frac{dw_y}{ds} = -\frac{p_0}{p_s}\frac{qB_0}{2p_0}\left(w_y + \frac{qB_0}{2p_0}x\right) + \mathcal{O}(2). \tag{3.101}$$

For the longitudinal, Eq. (3.88) to first order is

$$\begin{aligned}
\frac{dt}{ds} &= \left(-\frac{1}{\beta_0} - \Delta w_t\right)\left(1 - \frac{\Delta w_t}{\beta_0} + \cdots\right) \\
&= -\frac{1}{\beta_0} - \left(1 - \frac{1}{\beta_0^2}\right)\Delta w_t + \cdots \\
&= -\frac{1}{\beta_0} + \frac{\Delta w_t}{\gamma_0^2\beta_0^2} + \cdots.
\end{aligned} \tag{3.102}$$

As in CM: Chapter 3, we must calculate our time (or longitudinal position variable) relative to the design (synchronous) particle, so we really should have an extra $+\beta_0^{-1}$ to cancel the first term of Eq. (3.102). In CM: §3.7, we actually used an ultrarelativistic approximation, so that the $\Delta w_t/(\gamma_0\beta_0)^2$ term was ignored rather than anticipating the deeper discussion of longitudinal motion in CM: Chapter 7. § 7.4 shows how this term leads to an additional term in the M_{56} element of linear transport matrices.

b) In the paraxial approximation $p_s \simeq p_0$, so we may write the first order equations as

$$\frac{dx}{ds} \simeq w_x + \frac{y}{2r}, \tag{3.103}$$

$$\frac{dw_x}{ds} \simeq -\frac{x}{4r^2} + \frac{w_y}{2r}, \tag{3.104}$$

$$\frac{dy}{ds} \simeq -\frac{x}{2r} + w_y, \tag{3.105}$$

$$\frac{dw_y}{ds} \simeq -\frac{w_x}{2r} - \frac{y}{4r^2}. \tag{3.106}$$

Writing this in matrix form, produces

$$\frac{d}{ds} \begin{pmatrix} x \\ w_x \\ y \\ w_y \end{pmatrix} = \frac{1}{r} \begin{pmatrix} 0 & r & \frac{1}{2} & 0 \\ -\frac{1}{4r} & 0 & 0 & \frac{1}{2} \\ -\frac{1}{2} & 0 & 0 & r \\ 0 & -\frac{1}{2} & -\frac{1}{4r} & 0 \end{pmatrix} \begin{pmatrix} x \\ w_x \\ y \\ w_y \end{pmatrix}, \tag{3.107}$$

Define

$$\mathbf{K} = \begin{pmatrix} 0 & r & \frac{1}{2} & 0 \\ -\frac{1}{4r} & 0 & 0 & \frac{1}{2} \\ -\frac{1}{2} & 0 & 0 & r \\ 0 & -\frac{1}{2} & -\frac{1}{4r} & 0 \end{pmatrix}. \tag{3.108}$$

Raising $\mathbf{K}$ to successive powers gives:

$$\mathbf{K}^2 = \begin{pmatrix} -\frac{1}{2} & 0 & 0 & r \\ 0 & -\frac{1}{2} & -\frac{1}{4r} & 0 \\ 0 & -r & -\frac{1}{2} & 0 \\ \frac{1}{4r} & 0 & 0 & -\frac{1}{2} \end{pmatrix}, \tag{3.109}$$

$$\mathbf{K}^3 = \begin{pmatrix} 0 & -r & -\frac{1}{2} & 0 \\ \frac{1}{4r} & 0 & 0 & -\frac{1}{2} \\ \frac{1}{2} & 0 & 0 & -r \\ 0 & -\frac{1}{2} & -\frac{1}{4r} & 0 \end{pmatrix} = -\mathbf{K}, \tag{3.110}$$

$$\mathbf{K}^4 = -\mathbf{K}^2, \tag{3.111}$$

$$\mathbf{K}^5 = -\mathbf{K}^3 = \mathbf{K}. \tag{3.112}$$

The matrix for the whole solenoid is then

$$\mathbf{M} = \lim_{n \to \infty} \left(\mathbf{I} + \frac{1}{r}\mathbf{K}\frac{l}{n} \right)^n = e^{\mathbf{K}\, l/r} = e^{\mathbf{K}\phi}$$

$$= \mathbf{I} + \mathbf{K}\phi + \frac{1}{2!}\mathbf{K}^2\phi^2 + \cdots$$

$$= \mathbf{I} + \mathbf{K}\left(\phi - \frac{1}{3!}\phi^3 + \frac{1}{5!}\phi^5 + \cdots \right) + \mathbf{K}^2\left(\frac{1}{2!}\phi^2 - \frac{1}{4!}\phi^4 + \cdots \right)$$

$$= \mathbf{I} + \mathbf{K}\sin\phi + \mathbf{K}^2(1 - \cos\phi). \tag{3.113}$$

Substituting in the explicit matrices for $\mathbf{K}$ and $\mathbf{K}^2$, we have

$$
\mathbf{M} = \begin{pmatrix}
\frac{1+\cos\phi}{2} & r\sin\phi & \frac{\sin\phi}{2} & r(1-\cos\phi) \\
-\frac{\sin\phi}{4r} & \frac{1+\cos\phi}{2} & -\frac{1-\cos\phi}{4r} & \frac{\sin\phi}{2} \\
-\frac{\sin\phi}{2} & -r(1-\cos\phi) & \frac{1+\cos\phi}{2} & r\sin\phi \\
\frac{1-\cos\phi}{4r} & -\frac{\sin\phi}{2} & -\frac{\sin\phi}{4r} & \frac{1+\cos\phi}{2}
\end{pmatrix}
$$

$$(3.114)$$

This method for the computation of a solenoid matrix may be compared with the different method given in CM: §4.5.

3.9 Problem 3-9: Matrix for a tall rectangular solenoid

Repeat the previous problem with the vector potential

$$\vec{A} = \begin{cases} (0, B_0 x, 0), & \text{if } 0 < s < l; \\ 0, & \text{otherwise.} \end{cases} \tag{3.115}$$

Note that this gives a constant field along the axis of the magnet just like the previous problem, but the resulting transfer matrix is quite different. Explain this difference. What does the magnet look like?

Solution:

a) We can follow the solution of the previous problem (3.8) except that Eqs. (3.91 and 3.92) are now

$$\frac{\partial A_x}{\partial y} = 0, \tag{3.116}$$

$$\frac{\partial A_y}{\partial x} = B_0. \tag{3.117}$$

Hamilton's equations to first order are then

$$\frac{dx}{ds} = \frac{p_0}{p_s} w_x \simeq w_x, \tag{3.118}$$

$$\frac{dw_x}{ds} = \frac{p_0}{p_s} \frac{qB_0}{p_0} \left(w_y - \frac{qB_0}{p_0} x \right) + \mathcal{O}(2) \simeq \frac{1}{r} \left(w_y - \frac{x}{r} \right), \tag{3.119}$$

$$\frac{dy}{ds} = \frac{p_0}{p_s} \left(w_y - \frac{qB_0}{p_0} x \right) \simeq w_y - \frac{x}{r}, \tag{3.120}$$

$$\frac{dw_y}{ds} = 0. \tag{3.121}$$

b) Writing the linearized Hamilton's equations in matrix form:

$$\frac{d}{ds} \begin{pmatrix} x \\ w_x \\ y \\ w_y \end{pmatrix} = -\frac{1}{r} \mathbf{K} \begin{pmatrix} x \\ w_x \\ y \\ w_y \end{pmatrix}, \tag{3.122}$$

with

$$\mathbf{K} = \begin{pmatrix} 0 & r & 0 & 0 \\ -\frac{1}{r} & 0 & 0 & 1 \\ -1 & 0 & 0 & r \\ 0 & 0 & 0 & 0 \end{pmatrix}. \tag{3.123}$$

The sequence of powers of $\mathbf{K}$ are

$$\mathbf{K}^2 = \begin{pmatrix} -1 & 0 & 0 & r \\ 0 & -1 & 0 & 0 \\ 0 & -r & 0 & 0 \\ 0 & 0 & 0 & 0 \end{pmatrix}, \tag{3.124}$$

$$\mathbf{K}^3 = \begin{pmatrix} 0 & -r & 0 & 0 \\ \frac{1}{r} & 0 & 0 & -1 \\ 1 & 0 & 0 & -r \\ 0 & 0 & 0 & 0 \end{pmatrix} = -\mathbf{K}, \tag{3.125}$$

$$\mathbf{K}^4 = -\mathbf{K}^2, \tag{3.126}$$

$$\mathbf{K}^5 = \mathbf{K}. \tag{3.127}$$

Again, we have

$$\mathbf{M} = e^{\mathbf{K}\phi}$$

$$= \mathbf{I} + \mathbf{K}\left(\phi - \frac{1}{3!}\phi^3 + \frac{1}{5!}\phi^5 + \cdots\right) + \mathbf{K}^2\left(\frac{1}{2!}\phi^2 - \frac{1}{4!}\phi^4 + \cdots\right)$$

$$= \mathbf{I} + \mathbf{K}\sin\phi + \mathbf{K}^2(1 - \cos\phi), \tag{3.128}$$

which upon substitution of the matrices yields:

$$\mathbf{M} = \begin{pmatrix} \cos\phi & r\sin\phi & 0 & r(1-\cos\phi) \\ -\frac{\sin\phi}{r} & \cos\phi & 0 & \sin\phi \\ -\sin\phi & -r(1-\cos\phi) & 1 & r\sin\phi \\ 0 & 0 & 0 & 1 \end{pmatrix} \tag{3.129}$$

While the magnetic field inside the magnets of both problems 3–8 and 3–9 are identical, the fringe fields are not. Eq. (3.115) suggests a tall, thin solenoid as illustrated by Fig. 3.2, whereas Eq. (3.81) is the vector potential for a cylindrical solenoid.

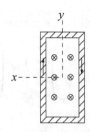

Fig. 3.2 Beam's-eye view of a tall, thin rectangular solenoid. The hashed area represent a coil with arrows indicating the direction of current flow. The magnetic field is into the page.

3.10 Problem 3–10: Zassenhaus formula

Verify Eq. (CM: 3.120) to second order in h.

$$e^{(\mathbf{A}+\mathbf{B})h} = e^{\mathbf{A}h/2} e^{\mathbf{B}h} e^{\mathbf{A}h/2} + \mathcal{O}(h^3). \qquad \text{(CM : 3.120)}$$

Solution:

Expanding the exponential on the left side of (CM: 3.120) gives

$$e^{(\mathbf{A}+\mathbf{B})h} = \mathbf{I} + (\mathbf{A} + \mathbf{B})h + \frac{1}{2}(\mathbf{A} + \mathbf{B})^2 h^2 + \cdots$$

$$= \mathbf{I} + (\mathbf{A} + \mathbf{B})h + (\mathbf{A}^2 + \mathbf{AB} + \mathbf{BA} + \mathbf{B}^2)\frac{h^2}{2} + \cdots. \qquad (3.130)$$

Expanding the right side gives

$$e^{\mathbf{A}h/2} e^{\mathbf{B}h} e^{\mathbf{A}h/2} = \left(\mathbf{I} + \mathbf{A}\frac{h}{2} + \mathbf{A}^2\frac{h^2}{8} + \cdots\right)\left(\mathbf{I} + \mathbf{B}h + \mathbf{B}^2\frac{h^2}{2} + \cdots\right)$$

$$\times \left(\mathbf{I} + \mathbf{A}\frac{h}{2} + \mathbf{A}^2\frac{h^2}{8} + \cdots\right)$$

$$= \left[\mathbf{I} + h\left(\frac{\mathbf{A}}{2} + \mathbf{B}\right) + h^2\left(\frac{\mathbf{AB}}{2} + \frac{\mathbf{A}^2}{8} + \frac{\mathbf{B}^2}{2}\right) + \cdots\right]$$

$$\times \left(\mathbf{I} + h\frac{\mathbf{A}}{2} + h^2\frac{\mathbf{A}^2}{8} + \cdots\right)$$

$$= \mathbf{I} + h\,(\mathbf{A} + \mathbf{B})$$

$$+ h^2\left(\frac{\mathbf{A}^2}{4} + \frac{\mathbf{BA}}{2} + \frac{\mathbf{AB}}{2} + \frac{\mathbf{A}^2}{8} + \frac{\mathbf{B}^2}{2} + \frac{\mathbf{A}^2}{8}\right) + \cdots$$

$$= \mathbf{I} + (\mathbf{A} + \mathbf{B})\,h$$

$$+ \left(\frac{\mathbf{A}^2}{2} + \frac{\mathbf{BA}}{2} + \frac{\mathbf{AB}}{2} + \frac{\mathbf{B}^2}{2}\right) h^2 + \cdots,$$

$$(3.131)$$

which is equal to the right side of Eq. (3.130) up to second order.

3.11   Problem 3–11: Cayley factorization

Show that[31, 46]

$$e^{\mathbf{SC}} = [\mathbf{I} + \tanh(\mathbf{SC}/2)]\,[\mathbf{I} - \tanh(\mathbf{SC}/2)]^{-1}, \tag{3.132}$$

where $\mathbf{C}$ is defined by Eq. (CM:3.108):

$$C_{jk} = \frac{\partial^2 H}{\partial X_j \partial X_k}(\widehat{X}). \tag{CM : 3.108}$$

Compare this with Eq. (3.147)

$$\mathbf{M} = (\mathbf{I} + \mathbf{SW})(\mathbf{I} - \mathbf{SW})^{-1}. \tag{CM : 3.147}$$

Solution:

For a real argument $x \in \mathbb{R}$, we have

$$\frac{1 + \tanh(x/2)}{1 - \tanh(x/2)} = \frac{1 + \frac{e^{x/2} - e^{-x/2}}{e^{x/2} + e^{-x/2}}}{1 - \frac{e^{x/2} - e^{-x/2}}{e^{x/2} + e^{-x/2}}} = \frac{e^{x/2} + e^{-x/2} + e^{x/2} - e^{-x/2}}{e^{x/2} + e^{-x/2} - (e^{x/2} - e^{-x/2})}$$

$$= \frac{2e^{x/2}}{2e^{-x/2}} = e^x. \tag{3.133}$$

Provided that $\tanh(\mathbf{SC}/2)$ is finite and that $[\mathbf{I} - \tanh(\mathbf{SC}/2)]^{-1}$ exists, then Eq. (3.132) will hold.

Comparison of Eqs. (3.132 and CM:3.147) indicates that

$$\mathbf{SW} = \tanh(\mathbf{SC}/2), \tag{3.134}$$

or, upon multiplication by $\mathbf{S}^{\mathrm{T}}$ from the left, that

$$\mathbf{W} = -\mathbf{S}\tanh(\mathbf{SC}/2). \tag{3.135}$$

This gives a direct path from the second derivatives of the Hamiltonian to the symmetric matrix $\mathbf{W}$ of the "Cayley" factorization of a transport matrix $\mathbf{M} = e^{\mathbf{SC}}$. Note that both $\mathbf{C}$ and $\mathbf{W}$ are symmetric matrices.

Chapter 4

Problems of Chapter 4: Optical Elements and Static Fields

4.1 Problem 4–1: Lithium lens

A lithium lens of length, l, and radius, a, has a current, I, flowing through it with a uniform current density. Consider a beam of antiprotons with momentum, p. What is the focal length of this lens?

Solution:

At a generic distance r from the lens axis the magnetic induction generated by this configuration is

$$B(r) = \mu_0 H(r) = \mu_0 \frac{I(r)}{2\pi r} = \mu_0 \frac{J\pi r^2}{2\pi r} = \frac{1}{2}\mu_0 Jr, \qquad (4.1)$$

where

$$J = \frac{I}{\pi a^2} \qquad (4.2)$$

is the uniform current density. Setting the direction of the current flow in the due way, i.e. opposite to the direction of the antiproton beam, we obtain a ring of Lorentz forces which squeeze the beam toward the lens axis. At a generic radius r these forces are all equal each other and their value is

$$F(r) = -evB(r) = -\frac{1}{2}ev\mu_0 Jr. \qquad (4.3)$$

Considering that we can write

$$\frac{d}{dt}\left(\gamma m \frac{dr}{dt}\right) = \gamma m \frac{d^2 r}{dt^2} \simeq \gamma m v^2 \frac{d^2 r}{dz^2}, \qquad (4.4)$$

47

since the Lorentz force does not affect the particle kinetic energy and having taken into account the paraxial approximation, we shall obtain the following equation

$$\frac{d^2r}{dt^2} + \left(\frac{e\mu_0 J}{2p}\right)r = \frac{d^2r}{dt^2} + K^2r = 0,\tag{4.5}$$

whose solutions are

$$r = A\cos(Kz) + B\sin(Kz),\tag{4.6}$$

$$\frac{dr}{dz} = K[-A\sin(Kz) + B\cos(Kz)],\tag{4.7}$$

where the coefficients A and B depend on the initial conditions

$$r(z = 0) = r_0, \qquad \text{and} \qquad \left(\frac{dr}{dz}\right)_{z=0} = 0,\tag{4.8}$$

as is typical any time that we are dealing with a problem of searching for a focal length. Hence we shall have $A = r_0$ and $B = 0$, so that

$$r = r_0\cos(Kz),\tag{4.9}$$

$$\frac{dr}{dz} = -Kr_0\sin(Kz).\tag{4.10}$$

At the end of the lens $(z = l)$, these last two equations become

$$r = r_0\cos(Kl),\tag{4.11}$$

$$\frac{dr}{dz} = -Kr_0\sin(Kl),\tag{4.12}$$

and the particle trajectory becomes the equation for a straight line:

$$r(z) = [r_0\cos(Kl)] - [Kr_0\sin(Kl)]z.\tag{4.13}$$

The focal length f is the value of z for which $r(z) = 0$:

$$f = \frac{\cot(Kl)}{K},\tag{4.14}$$

with

$$K = \sqrt{\frac{e\mu_0 J}{2p}}.\tag{4.15}$$

4.2 Problem 4–2: Multipoles from scalar potential

Show that Eqs. (CM: 4.11 and CM: 4.22):

$$\vec{H} = \vec{\nabla}\Psi, \qquad \text{(CM: 4.11)}$$

$$\Psi = \sum_{n=0}^{\infty} \frac{a}{n+1} \left(\frac{r}{a}\right)^{n+1} \left[F_n \cos\left((n+1)\theta\right) + G_n \sin\left((n+1)\theta\right)\right], \quad \text{(CM: 4.22)}$$

yield field components of the forms given in Eqs. (CM:4.6 and CM:4.7):

$$B_y = B_0 \sum_{n=0}^{\infty} \left(\frac{r}{a}\right)^n \left(b_n \cos n\theta - a_n \sin n\theta\right), \qquad \text{(CM: 4.6)}$$

$$B_x = B_0 \sum_{n=0}^{\infty} \left(\frac{r}{a}\right)^n \left(a_n \cos n\theta + b_n \sin n\theta\right), \qquad \text{(CM: 4.7)}$$

which in turn satisfy Maxwell's equations.

Solution:

In polar coordinates, we may evaluate

$$B_r = \mu_0 \frac{\partial \Psi}{\partial r} = \mu_0 \sum_{n=0}^{\infty} \left(\frac{r}{a}\right)^n \left[F_n \cos\left((n+1)\theta\right) + G_n \sin\left((n+1)\theta\right)\right], \quad (4.16)$$

$$B_\theta = \frac{\mu_0}{r} \frac{\partial \Psi}{\partial \theta} = \mu_0 \sum_{n=0}^{\infty} \left(\frac{r}{a}\right)^n \left[G_n \cos\left((n+1)\theta\right) - F_n \sin\left((n+1)\theta\right)\right].$$

$$(4.17)$$

These two equations can be written in matrix form as

$$\begin{pmatrix} B_r \\ B_\theta \end{pmatrix} = \mu_0 \sum_{n=0}^{\infty} \left(\frac{r}{a}\right)^n \begin{pmatrix} \cos\left((n+1)\theta\right) & \sin\left((n+1)\theta\right) \\ -\sin\left((n+1)\theta\right) & \cos\left((n+1)\theta\right) \end{pmatrix} \begin{pmatrix} F_n \\ G_n \end{pmatrix}. \quad (4.18)$$

Converting to Cartesian coordinates this becomes just a simple rotation:

$$\begin{pmatrix} B_x \\ \hat{B}_y \end{pmatrix} = \begin{pmatrix} \cos\theta & -\sin\theta \\ \sin\theta & \cos\theta \end{pmatrix} \begin{pmatrix} B_r \\ B_\theta \end{pmatrix}$$

$$= B_0 \sum_{n=0}^{\infty} \left(\frac{r}{a}\right)^n \begin{pmatrix} \cos(n\theta) & \sin(n\theta) \\ -\sin(n\theta) & \cos(n\theta) \end{pmatrix} \begin{pmatrix} a_n \\ b_n \end{pmatrix}, \quad (4.19)$$

which is equivalent to Eqs. (CM: 4.7 and CM: 4.6) since we have $G_n = B_0 b_n / \mu_0$ and $F = B_0 a_n / \mu_0$ as stated in the CM text following Eq. (CM: 4.12).

Next we need to verify that Eqs. (CM: 4.6 and CM: 4.7) satisfy Maxwell's equations:

$$\nabla \cdot \vec{B} = 0, \tag{4.20}$$

$$\nabla \times \vec{B} = 0. \tag{4.21}$$

Since B_x and B_y do not depend on z and also $B_z = 0$, the only nontrivial part of Eq. (4.21) will be demonstrating that

$$\frac{\partial B_y}{\partial x} - \frac{\partial B_x}{\partial y} = 0. \tag{4.22}$$

Likewise for Eq. (4.20) we just need to show that

$$\frac{\partial B_x}{\partial x} + \frac{\partial B_y}{\partial y} = 0. \tag{4.23}$$

Starting with the field components in matrix form Eq. (4.19), we can write

$$\begin{pmatrix} B_x \\ B_y \end{pmatrix} = B_0 \sum_{n=0}^{\infty} \frac{1}{a^n} \begin{pmatrix} r\cos\theta & r\sin\theta \\ -r\sin\theta & r\cos\theta \end{pmatrix}^n \begin{pmatrix} a_n \\ b_n \end{pmatrix},$$

$$= B_0 \sum_{n=0}^{\infty} \frac{1}{a^n} \begin{pmatrix} x & y \\ -y & x \end{pmatrix}^n \begin{pmatrix} a_n \\ b_n \end{pmatrix}. \tag{4.24}$$

Rather than derive (or look up) the expansions for $\cos n\theta$ and $\sin n\theta$, we can use this matrix form to kill two birds with the same stone, since we may write:

$$\begin{pmatrix} \nabla \cdot \vec{B} \\ (\nabla \times \vec{B})_z \end{pmatrix} = \begin{pmatrix} \frac{\partial B_x}{\partial x} + \frac{\partial B_y}{\partial y} \\ \frac{\partial B_y}{\partial x} - \frac{\partial B_x}{\partial y} \end{pmatrix}$$

$$= \frac{\partial}{\partial x}\begin{pmatrix} B_x \\ B_y \end{pmatrix} + \begin{pmatrix} 0 & 1 \\ -1 & 0 \end{pmatrix}\frac{\partial}{\partial y}\begin{pmatrix} B_x \\ B_y \end{pmatrix}, \tag{4.25}$$

with

$$\frac{\partial}{\partial x}\begin{pmatrix} B_x \\ B_y \end{pmatrix} = B_0 \sum_{n=1}^{\infty} \frac{n}{a^n} \begin{pmatrix} x & y \\ -y & x \end{pmatrix}^{n-1} \begin{pmatrix} 1 & 0 \\ 0 & 1 \end{pmatrix} \begin{pmatrix} a_n \\ b_n \end{pmatrix}$$

$$= B_0 \sum_{n=1}^{\infty} \frac{nr^n}{a^n} \begin{pmatrix} C & S \\ -S & C \end{pmatrix} \begin{pmatrix} a_n \\ b_n \end{pmatrix}, \tag{4.26}$$

$$\frac{\partial}{\partial y}\begin{pmatrix} B_x \\ B_y \end{pmatrix} = B_0 \sum_{n=1}^{\infty} \frac{nr^n}{a^n} \begin{pmatrix} C & S \\ -S & C \end{pmatrix} \begin{pmatrix} 0 & 1 \\ -1 & 0 \end{pmatrix} \begin{pmatrix} a_n \\ b_n \end{pmatrix}, \tag{4.27}$$

where $C = \cos[(n-1)\theta]$ and $S = \sin[(n-1)\theta]$. Comparing of the last three equations, all we need to demonstrate is

$$\begin{pmatrix} C & S \\ -S & C \end{pmatrix} + \begin{pmatrix} 0 & 1 \\ -1 & 0 \end{pmatrix} \begin{pmatrix} C & S \\ -S & C \end{pmatrix} \begin{pmatrix} 0 & 1 \\ -1 & 0 \end{pmatrix}$$

$$= \begin{pmatrix} C & S \\ -S & C \end{pmatrix} + \begin{pmatrix} -S & C \\ -C & -S \end{pmatrix} \begin{pmatrix} 0 & 1 \\ -1 & 0 \end{pmatrix}$$

$$= \begin{pmatrix} C & S \\ -S & C \end{pmatrix} + \begin{pmatrix} -C & -S \\ S & -C \end{pmatrix}$$

$$= \begin{pmatrix} 0 & 0 \\ 0 & 0 \end{pmatrix}. \tag{4.28}$$

4.3   Problem 4–3: Dipole C-magnet

a) Assuming an infinite plane separates the boundary between two magnetic materials of permeability μ_1 and μ_2 respectively, use Maxwell's equations to calculate the change in both the normal and tangential components of both $\vec{H}$ and $\vec{B}$ across the boundary.

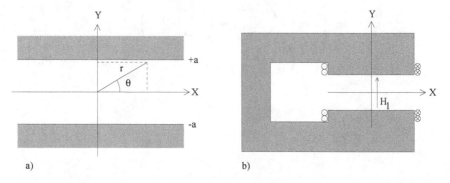

Fig. 4.1   a) Pole profile for a "dipole" magnet.   b) Dipole magnet with coils and dipole field.

b) Using these boundary conditions, estimate the magnetic field in the air gap of an iron dipole magnet like that illustrated in Fig. 4.1b, with an approximate path through the iron of $L = 30$ cm and a gap $g = 2$ cm. There are a total of 100 turns of wire carrying 100 A. Assume that $\mu/\mu_0 = 1000$. What is the magnitude of B? What is the radius of curvature of a proton with a kinetic energy of 600 MeV?

c) How would this change if $\mu/\mu_0 = 20$?

Solution:

a) Applying Gauss' law for magnetic fields to a thin region parallel to the boundary of a surface as illustrated in Fig. 4.2a, we have

$$\int_{\partial V} \vec{B} \cdot d\vec{S} = \int_V \nabla \cdot \vec{B} \, dV = 0, \tag{4.29}$$

$$0 = A(\vec{B}_2 - \vec{B}_1) \cdot \hat{n}, \tag{4.30}$$

which gives $B_{\perp 1} = B_{\perp 2}$. Since $\vec{B} = \mu \vec{H}$, this also yields

$$\frac{H_{\perp 2}}{\mu_2} = \frac{H_{\perp 1}}{\mu_1}, \tag{4.31}$$

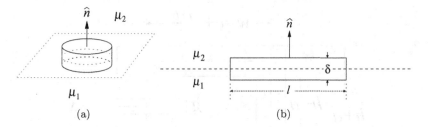

Fig. 4.2   a) A thin pillbox Gaussian volume crossing the boundary between two different materials. b) Path for a line integral of $\vec{H}$ at a boundary between two regions.

at the boundary.

Integrating around the thin loop shown in Fig. 4.2b in a clockwise direction we have

$$\int_A \vec{J} \cdot d\vec{S} = \int_A (\nabla \times \vec{H}) \cdot d\vec{S} = \int_{\partial A} \vec{H} \cdot d\vec{l}$$

$$= (H_{\|2} - H_{\|1})l + \left[\left(\frac{B_{\perp 2}}{\mu_2} + \frac{B_{\perp 1}}{\mu_1} \right)_{\substack{\text{left}\\\text{end}}} - \left(\frac{B_{\perp 2}}{\mu_2} + \frac{B_{\perp 1}}{\mu_1} \right)_{\substack{\text{right}\\\text{end}}} \right] \frac{\delta}{2}$$

$$\to (H_{\|2} - H_{\|1})l \quad \text{as } \delta \to 0, \tag{4.32}$$

where a positive surface current along the boundary would be coming out of the pages. Upon reordering, we see that the difference of $H_\|$ from one side of the boundary to the other to the other side is equal to any surface current flowing along the boundary through the loop:

$$(H_{\|2} - H_{\|1}) = \frac{1}{l} \int_A \vec{J} \cdot d\vec{S}. \tag{4.33}$$

If there is no current flowing at the boundary, then $H_\|$ is continuous across the boundary. (See any standard E&M text[34] for more discussion of these conditions.)

b) Estimating the line integral of $\vec{H}$ around a closed loop through the iron (see Fig. 4.3), we have

$$NI \simeq \frac{B}{\mu_0} \left(\frac{L}{\mu_r} + g \right),$$

where $\mu_r = \mu/\mu_0$, or

$$B = \frac{4\pi \times 10^{-7} \ [\text{Tm/A}] \times 100 \times 100 \ [\text{A}]}{(0.3/1000 + 0.02) \ [\text{m}]} \simeq 0.62 \ [\text{T}]. \tag{4.34}$$

A proton of $K = 600$ MeV kinetic energy has the magnetic rigidity

$$\frac{p}{q} = mc\sqrt{\left(1 + \frac{K}{mc^2} \right)^2 - 1} = 4.066 \text{ Tm}. \tag{4.35}$$

So the bending radius will about $\rho = p/(qB) \simeq 6.6$ m.

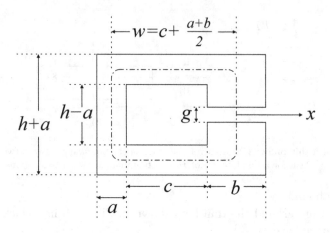

Fig. 4.3 Approximation to the path (dash-dot curve) of the line integral in part b. For $x = 0$, the total path length through the iron is approximately $L_0 = L = 2(w + h) - g$.

c) If the relative permeability is reduced to 20, then the value estimated in Eq. 4.34 should be reduced roughly by a factor of

$$\frac{(0.3/1000 + 0.02)[\text{m}]}{(0.3/20 + 0.02)[\text{m}]} = \frac{0.0203}{0.035} = 0.58, \qquad (4.36)$$

which lowers the field in the gap to $B \simeq 0.36$ T. The bending radius would increase to $\rho \simeq 11.3$ m.

We should also expect a larger quadrupole component to the field in the gap, since the gap field would be more sensitive to the path length through the "iron".

Comment:

To estimate the gradient with a constant μ_r, we can look at the variation of L as a function of x position. The magnetic flux lines do not cross, so we can expect that as the point of interest moves in the $+x$ direction, the path of integration along a given flux line will increase. If we assume the path length increase in iron is approximately linear in x, then we can write for $x = ba/2$

$$L(\alpha) \simeq 2(h + a\alpha) + 2\left(w + \frac{a+b}{2}\alpha\right) - g, \qquad (4.37)$$

$$L(x) = 2(h + w) - g + 2\frac{3a + b}{b}x = L_0 + 2\left(1 + \frac{3a}{b}\right)x, \qquad (4.38)$$

and

$$B_y(x) \simeq \frac{\mu_0 N I}{\frac{L_0}{\mu_r} + g + \frac{2}{\mu_r}\left(1 + \frac{3a}{b}\right)x}. \tag{4.39}$$

Differentiating and evaluating at $x = 0$ gives

$$\frac{\partial B_y}{\partial x} \simeq -\frac{\beta}{L_0} B_y(0), \quad \text{with} \quad \beta \simeq \frac{2}{\mu_r}\left(1 + \frac{3a}{b}\right). \tag{4.40}$$

Of course for accurate calculations, the permeability varies throughout the iron, and something like a finite-difference or finite element-code should be used to calculate the real field distribution in the gap.

4.4   Problem 4–4: Nonlinear drift equations

Using the coordinates x, $x' = dx/ds$, y, $y' = dy/ds$, $z = -v\,\Delta t$, and δ, (Δt is the deviation in the time that a particle passes the location s from the time that the design particle passes the same location. In other words, z is the path length difference of the particle in question from the design particle.) Find an exact transformation function for a trajectory through a drift space of length l. Note that the z–component is nonlinear.

Solution:

First we should note that (x, x') and (y, y') are not exact canonical pairs; however the statement of the problem does not make that claim.

Start $s = 0$ from the position $(x_0, x'_0, y_0, y'_0, z_0, \delta_0)$.

For the components we have simply linear expressions for all but the longitudinal coordinate:

$$x(s) = x_0 + x'_0 s, \tag{4.41}$$

$$x'(s) = x'_0, \tag{4.42}$$

$$y(s) = y_0 + y'_0 s, \tag{4.43}$$

$$y'(s) = y'_0, \tag{4.44}$$

$$\delta(s) = \delta_0. \tag{4.45}$$

For z we need to use Pythagoras's theorem to calculate the path length of the trajectory from the beginning to end of the drift:

$$
\begin{aligned}
l(s) &= \sqrt{[x(s) - x_0]^2 + [y(s) - y_0]^2 + [s - 0]^2} \\
&= s\sqrt{1 + x'^2_0 + y'^2_0}.
\end{aligned}
\tag{4.46}
$$

The excess path length to the end of the drift is then

$$z(s) = -[l(s) - s] = s\left[1 - \sqrt{1 + x'^2_0 + y'^2_0}\right]. \tag{4.47}$$

Expanding this, we find to forth order in the slopes,

$$z(s) = -s\left[\frac{x'^2_0 + y'^2_0}{2} - \frac{(x'^2_0 + y'^2_0)^2}{8} + \cdots\right]. \tag{4.48}$$

Comment:

As mentioned earlier, x' and y' are not canonical momenta of the coordinates x and y, respectively. If we follow the discussion of CM: § 3.6 with a Hamiltonian $H(x, w_x, y, w_y, z, \delta; s)$, where the canonical momenta are

$$w_x = \frac{P_x}{p_0} = \frac{p_x}{p_0} + \frac{q}{p_0} A_x,$$ (4.49)

$$w_y = \frac{P_y}{p_0} = \frac{p_y}{p_0} + \frac{q}{p_0} A_y,$$ (4.50)

$$\delta = \frac{\Delta p}{p_0} = \frac{\Delta U}{\beta_0 c\, p_0}$$ (4.51)

with the longitudinal coordinate given by $z = -\beta_0 c \Delta t$, we can find exact expressions for the transverse canonical momenta in terms of the slopes

$$x' = \frac{p_x}{p_z}, \quad \text{and} \quad y' = \frac{p_y}{p_z},$$ (4.52)

as follows. The total momentum may be written as

$$p = p_0(1 + \delta) = p_z \sqrt{1 + x'^2 + y'^2}.$$ (4.53)

Solving for p_0 we obtain

$$p_0 = \frac{p_z \sqrt{1 + x'^2 + y'^2}}{1 + \delta},$$ (4.54)

which may be substituted back into Eqs. (4.49 and 4.50) for the desired results:

$$w_x = \frac{(1 + \delta)x'}{\sqrt{1 + x'^2 + y'^2}} + \frac{q}{p_0} A_x,$$ (4.55)

$$w_y = \frac{(1 + \delta)y'}{\sqrt{1 + x'^2 + y'^2}} + \frac{q}{p_0} A_y.$$ (4.56)

We can also write the slopes in terms of the canonical momenta as

$$x' = \frac{w_x - \frac{qA_x}{p_0}}{\sqrt{(1 + \delta)^2 - (w_x - \frac{qA_x}{p_0})^2 - (w_y - \frac{qA_y}{p_0})^2}},$$ (4.57)

$$y' = \frac{w_y - \frac{qA_y}{p_0}}{\sqrt{(1 + \delta)^2 - (w_y - \frac{qA_y}{p_0})^2 - (w_y - \frac{qA_y}{p_0})^2}}.$$ (4.58)

4.5   Problem 4–5: Path length through thick-lens quad

Find a formula for the excess path length of a trajectory passing through a thick quadrupole lens element.

Solution:

The excess path length along the trajectory is

$$\Delta l = l(s) - s = \int_0^s \sqrt{1 + x'^2 + y'^2}\, ds - s,$$

$$= \frac{1}{2} \int_0^s (x'^2 + y'^2)\, ds + \mathcal{O}(4), \tag{4.59}$$

to second order in the slopes:

$$x' = -x_0 \sqrt{k} \sin(\sqrt{k}s) + x_0' \cos(\sqrt{k}s), \tag{4.60}$$

$$y' = y_0 \sqrt{k} \sinh(\sqrt{k}s) + y_0' \cosh(\sqrt{k}s). \tag{4.61}$$

Here we have assumed a horizontally focusing quadrupole. The squares of these slopes are

$$x'^2 = x_0^2 k \frac{1 - \cos(2\sqrt{k}s)}{2} + x_0'^2 \frac{1 + \cos(2\sqrt{k}s)}{2} - x_0 x_0' \sqrt{k}\, \sin(2\sqrt{k}s),$$
$$\tag{4.62}$$

$$y'^2 \doteq y_0^2 k \frac{\cosh(2\sqrt{k}s) - 1}{2} + y_0'^2 \frac{\cosh(2\sqrt{k}s) + 1}{2} + y_0 y_0' \sqrt{k}\, \sinh(2\sqrt{k}s).$$
$$\tag{4.63}$$

Integrating Eq. (4.62) yields

$$\int_0^s x'^2\, ds = \frac{(x_0^2 k + x_0'^2)s}{2} + \frac{(x_0'^2 - kx_0^2)}{4\sqrt{k}} \sin(2\sqrt{k}s) - \frac{x_0 x_0'}{2} \cos(2\sqrt{k}s)$$
$$\tag{4.64}$$

Similarly integration of Eq. (4.63) gives

$$\int_0^s y'^2\, ds = \frac{(y_0'^2 - y_0^2 k)s}{2} + \frac{(y_0'^2 + ky_0^2)}{4\sqrt{k}} \sinh(2\sqrt{k}s) + \frac{y_0 y_0'}{2} \cosh(2\sqrt{k}s).$$
$$\tag{4.65}$$

Adding these last two equations and dividing by two produces the second-order formula for the excess path length:

$$\Delta l \simeq \frac{1}{2} \left[\frac{(x_0^2 - y_0^2)k + x_0'^2 + y_0'^2}{2} s + \frac{(x_0'^2 - kx_0^2)}{4\sqrt{k}} \sin(2\sqrt{k}s) \right.$$
$$+ \frac{(y_0'^2 + ky_0^2)}{4\sqrt{k}} \sinh(2\sqrt{k}s) - \frac{x_0 x_0'}{2} \cos(2\sqrt{k}s)$$
$$\left. + \frac{y_0 y_0'}{2} \cosh(2\sqrt{k}s) \right]. \tag{4.66}$$

4.6　Problem 4–6: Sector magnet transport functions

Ignoring fringe field effects, find exact expressions for the transformation of a particle through a sector magnet with ends perpendicular to the design particle trajectory.

Solution:

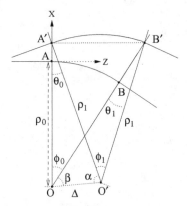

Fig. 4.4　Projection of the test and design particles' orbits in the horizontal plane. The design orbit enters the magnet at A with transverse coordinates $(x, x', y, y') = (0, 0, 0, 0)$ and leaves at B with transverse coordinates $(x, x', y, y') = (0, 0, 0, 0)$. A test particle enters at A′ with initial coordinates (x_0, x_0', y_0, y_0') and leaves at B′ with coordinates (x_1, x_1', y_1, y_1'). The design trajectory bends by an angle ϕ_0 with a radius $\rho_0 = \overline{OA} = \overline{OB}$, whereas the test particle bends through the angle ϕ_1 with radius $\rho_1 = \overline{O'A'} = \overline{O'B'}$. The distance between the centers of the two arcs is $\Delta = \overline{OO'}$.

The radius of curvature of the design orbit is

$$\rho_0 = \frac{p_0}{qB_0}, \tag{4.67}$$

so the point O is located at $(-\rho_0, 0)$ in the Cartesian XZ-plane. For the test particle entering the magnet at point A, we have the momentum

$$p = p_0(1 + \delta), \tag{4.68}$$

with the components:

$$p_{xi} = \frac{px_0'}{\sqrt{1 + x_0'^2 + y_0'^2}}, \quad p_{yi} = \frac{py_0'}{\sqrt{1 + x_0'^2 + y_0'^2}}, \quad p_{zi} = \frac{p}{\sqrt{1 + x_0'^2 + y_0'^2}}. \tag{4.69}$$

Projected into the horizontal plane, the radius of curvature of the test particle must then be

$$\rho_1 = \rho_0(1 + \delta) \sqrt{\frac{1 + x_0'^2}{1 + x_0'^2 + y_0'^2}}. \tag{4.70}$$

The angle θ_0 of Fig 4.4 is given by

$$\theta_0 = \tan^{-1} x_0', \tag{4.71}$$

and we find that the location of point O' in the XZ-plane is

$$O': \quad (x_0 - \rho_1 \cos \theta_0, \quad \rho_1 \sin \theta_0). \tag{4.72}$$

The equation of the line $\overline{OB'}$ may be written as

$$X = Z \cot \phi_0 - \rho_0. \tag{4.73}$$

The circle of radius ρ_1 centered at O' is given by

$$\rho_1^2 = [X - (x_0 - \rho_1 \cos \theta_0)]^2 + [Z - \rho_1 \sin \theta_0]^2. \tag{4.74}$$

The coordinates $(X_{B'}, Z_{B'})$ of the point B' will be located at the intersection of this circle (4.74) with the line (4.73).

Now, combining Eqs. (4.73 and 4.74) to solve for $Z_{B'}$:

$$\rho_1^2 = [Z_{B'} \cot \phi_0 - (\rho_0 + x_0) + \rho_1 \cos \theta_0]^2 + [Z_{B'} - \rho_1 \sin \theta_0]^2; \tag{4.75}$$

$$\begin{aligned}
0 &= \frac{Z_{B'}^2}{\sin^2 \phi_0} - 2Z_{B'}[\rho_1 \sin \theta_0 + (\rho_0 + x_0 - \rho_1 \cos \theta_0) \cot \phi_0] \\
&\quad + (\rho_0 + x_0)^2 - 2(\rho_0 + x_0)\rho_1 \cos \theta_0 \\
&= Z_{B'}^2 - 2Z_{B'}[(\rho_0 + x_0) \cos \phi_0 - \rho_1 \cos(\phi_0 + \theta_0)] \sin \phi_0 \\
&\quad + [(\rho_0 + x_0)^2 - 2(\rho_0 + x_0)\rho_1 \cos \theta_0] \sin^2 \phi_0. \\
&= Z_{B'}^2 - 2b \sin \phi_0 \, Z_{B'} + c \sin^2 \phi_0, \tag{4.76}
\end{aligned}$$

where

$$b = [(\rho_0 + x_0) \cos \phi_0 - \rho_1 \cos(\phi_0 + \theta_0)], \tag{4.77}$$

$$c = [(\rho_0 + x_0)^2 - 2(\rho_0 + x_0)\rho_1 \cos \theta_0], \tag{4.78}$$

so we have

$$Z_{B'} = \left(b \pm \sqrt{b^2 - c}\right) \sin \phi_0. \tag{4.79}$$

We will restrict the magnet to have $\phi_0 \leq 180°$, so we should expect that $Z_{B'} > 0$, provided that the particle exits from the downstream edge of the sector. If the $b^2 - c < 0$, then the trajectory will bend around inside the magnet to come back out the entrance face of the magnet (i. e. typically when the B field is too large for the particle's rigidity).

Let us examine the sign of

$$c = (\rho_0 + x_0)[\rho_0 + x_0 - 2\rho_1 \cos \theta_0]. \tag{4.80}$$

Except in extreme cases, we may expect that $|x_0| \ll \rho_0$, $\rho_1 \sim \rho_0$, and $|\theta_0| \ll \pi/2$, so it is likely that $c < 0$. If we assume that c is negative, then the solution of Eq. (4.79) should be

$$Z_{B'} = \left(b + \sqrt{b^2 - c}\right) \sin \phi_0, \tag{4.81}$$

and recalling Eq. (4.73)

$$X_{B'} = \left(b + \sqrt{b^2 - c}\right) \cos \phi_0 - \rho_0. \tag{4.82}$$

Of course a simple evaluation of Eq. (4.78) would verify if this assumption is justified.

The length of $\overline{A'B'}$ is

$$2\rho_1 \sin \frac{\phi_1}{2} = \sqrt{(X_{B'} - x_0)^2 + Z_{B'}^2}, \tag{4.83}$$

which when solved for ϕ_1 yields

$$\phi_1 = 2 \sin^{-1} \left(\frac{\sqrt{(X_{B'} - x_0)^2 + Z_{B'}^2}}{2\rho_1} \right). \tag{4.84}$$

The horizontal slope at the exit is

$$x_1' = \tan \theta_1 = \tan(\theta_0 + \phi_0 - \phi_1), \tag{4.85}$$

and the desired radial coordinate is

$$x_1 = \sqrt{Z_{B'}^2 + (X_{B'} + \rho_0)^2} - \rho_0. \tag{4.86}$$

Even though we have not yet calculated the vertical slope y_1', we may write:

$$p_{xf} = \frac{px_1'}{\sqrt{1 + x_1'^2 + y_1'^2}}, \tag{4.87}$$

$$p_{yf} = \frac{py_1'}{\sqrt{1 + x_1'^2 + y_1'^2}} = p_{yi} = \frac{py_0'}{\sqrt{1 + x_0'^2 + y_0'^2}}, \tag{4.88}$$

$$p_{zf} = \frac{p}{\sqrt{1 + x_1'^2 + y_1'^2}}, \tag{4.89}$$

but since the magnitude of the momentum must remain constant, we have

$$1 = \frac{1 + x_1'^2}{1 + x_1'^2 + y_1'^2} + \frac{y_0'^2}{1 + x_0'^2 + y_0'^2}. \tag{4.90}$$

Solving this for y_1' we have:

$$1 + \frac{y_1'^2}{1 + x_1'^2} = \frac{1}{1 - \frac{y_0'^2}{1 + x_0'^2 + y_0'^2}}; \tag{4.91}$$

$$1 + \frac{y_1'^2}{1 + x_1'^2} = 1 + \frac{y_0'^2}{1 + x_0'^2}; \tag{4.92}$$

$$y_1' = \sqrt{\frac{1 + x_1'^2}{1 + x_0'^2}} \, y_0'. \tag{4.93}$$

The projected path length in the XZ-plane is

$$l_h = \rho_1 \phi_1, \tag{4.94}$$

and the time of flight from A' to B' is $t = l_h/v_h$ where v_h is the horizontal velocity component (i. e., in the XZ-plane). So

$$y_1 = y_0 + v_y \frac{l_h}{v_h} = y_0 + \frac{\rho_1 \phi_1 y_0'}{\sqrt{1 + x_0'^2}}. \tag{4.95}$$

The total path length then is

$$l = \sqrt{l_h^2 + (y_1 - y_0)^2} = \rho_1 \phi_1 \sqrt{1 + \frac{y_0'^2}{1 + x_0'^2}} = \rho_1 \phi_1 \sqrt{\frac{1 + x_0'^2 + y_0'^2}{1 + x_0'^2}}, \tag{4.96}$$

and the excess path length relative to the design trajectory must be

$$\Delta l = \rho_1 \phi_1 \sqrt{\frac{1 + x_0'^2 + y_0'^2}{1 + x_0'^2}} - \rho_0 \phi_0. \tag{4.97}$$

so

$$z_1 = z_0 - \left(\rho_1 \phi_1 \sqrt{\frac{1 + x_0'^2 + y_0'^2}{1 + x_0'^2}} - \rho_0 \phi_0 \right). \tag{4.98}$$

The desired final variables $(x_1, x_1', y_1, y_1', z_1, \delta_1)$ are respectively given by Eqs. (4.86, 4.85, 4.95, 4.93, and 4.98) together with the trivial equation

$$\delta_1 = \delta_0. \tag{4.99}$$

4.7   Problem 4–7: Vector potential for multipole expansion

a) Assuming that $x/\rho \ll 1$, find an expression for the vector potential $\vec{A}$ which yields the field given by Eqs. (CM: 4.6 and CM: 4.7):

$$B_y = B_0 \sum_{n=0}^{\infty} \left(\frac{r}{a}\right)^n (b_n \cos n\theta - a_n \sin n\theta), \qquad \text{(CM: 4.6)}$$

$$B_x = B_0 \sum_{n=0}^{\infty} \left(\frac{r}{a}\right)^n (a_n \cos n\theta + b_n \sin n\theta). \qquad \text{(CM: 4.7)}$$

b) Find a first order correction in x/ρ to this.

Solution:

a) See Problem 4–15b for the solution. From Eq. (4.184), we have

$$A_s = B_0 \sum_{n=0}^{\infty} \frac{a}{n+1} \left(\frac{r}{a}\right)^{n+1} [a_n \sin((n+1)\theta) - b_n \cos((n+1)\theta)]. \quad (4.100)$$

Note that in plasma physics the coordinates (r, θ, ϕ) with $\phi = s/\rho$ are referred to as *simple toroidal coordinates* as opposed to the "toroidal coordinates" usually discussed in mathematical physics texts[7, 49].

b) Part a) assumed a Cartesian coordinate system, but here we need to look at what happens when we consider cylindrical coordinates for a slight curvature of the bend magnet. Perhaps this is not that well-defined a problem, so we must make some assumptions. (See, e. g. Ref. [48] for a more general treatment of fields in cylindrical coordinates.)

The inspiration for this problem came from the method used to construct magnets for the SSC (Superconducting Super Collider) that was never completed. For a large radius bend, the dipole magnets were first built as straight magnets and afterwards bent to the desired curvature. The technique for producing this curvature is to make a few welds along one side of the outside of the helium containment vessel of the cold mass. The contraction of the weld as it cools then gives a curve to the magnet.

Let us take the solution in Eq. (4.100) and multiply by the $1 + x/\rho$ factor of a slightly modified version of Eq. (CM: 3.44):

$$A_s = \left(1 + \frac{x}{\rho}\right) (\vec{A}_a \cdot \hat{\phi}), \qquad (4.101)$$

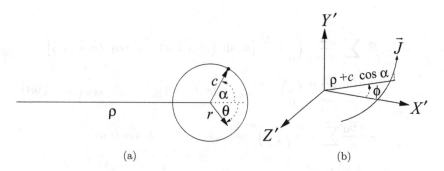

Fig. 4.5 a) The field at azimuth $\phi = 0$ is calculated for a point at polar coordinate (r, θ) due to a wire placed at (c, α). The center of the polar coordinate system is displaced by ρ along the X'-axis in a lab Cartesian system. b) Indication of the slightly bowed magnet in the 3-d lab coordinate system.

where we have replaced $\hat{\theta}$ by $\hat{\phi}$ to agree with the notation in Fig. 4.5, and $\vec{A}_a$ is the vector potential from Eq. (4.100), so we get

$$A_s = A_a + \frac{a}{\rho} \frac{r}{a} \cos \theta \, A_a. \tag{4.102}$$

The factor r/a in the second term suggests that we might expect a sort of "feed-up" of multipoles in which a given multipole leads to higher multipole components. We will use the simple identities:

$$\cos \theta \cos \left((n+1)\theta\right) = \frac{1}{2} \left[\cos \left((n+2)\theta\right) + \cos(n\theta)\right], \tag{4.103}$$

$$\cos \theta \sin \left((n+1)\theta\right) = \frac{1}{2} \left[\sin \left((n+2)\theta\right) + \sin(n\theta)\right]. \tag{4.104}$$

So Eq. (4.102) becomes

$$
\begin{aligned}
A_s = B_0 \sum_{n=0}^{\infty} \frac{a}{n+1} \left(\frac{r}{a}\right)^{n+1} & \left[a_n \sin \left((n+1)\theta\right) - b_n \cos \left((n+1)\theta\right)\right] \\
+ \frac{aB_0}{\rho} \sum_{n=0}^{\infty} \frac{a}{n+1} \left(\frac{r}{a}\right)^{n+2} & \left[a_n \sin \left((n+2)\theta\right) - b_n \cos \left((n+2)\theta\right)\right] \\
+ \frac{aB_0}{\rho} \sum_{n=0}^{\infty} \frac{a}{n+1} \left(\frac{r}{a}\right)^{n+2} & \left[a_n \sin \left(n\theta\right) - b_n \cos \left(n\theta\right)\right]
\end{aligned}
$$

$$= B_0 \sum_{n=0}^{\infty} \frac{a}{n+1} \left(\frac{r}{a}\right)^{n+1} \left[a_n \sin\left((n+1)\theta\right) - b_n \cos\left((n+1)\theta\right)\right]$$

$$+ \frac{aB_0}{\rho} \sum_{n=1}^{\infty} \frac{a}{n} \left(\frac{r}{a}\right)^{n+1} \left[a_{n-1} \sin\left((n+1)\theta\right) - b_{n-1} \cos\left((n+1)\theta\right)\right]$$

$$+ \frac{aB_0}{\rho} \sum_{n=0}^{\infty} \frac{a}{n+1} \left(\frac{r}{a}\right)^{n+2} \left[a_n \sin\left(n\theta\right) - b_n \cos\left(n\theta\right)\right]$$

$$= -B_0 b_0 r \cos\theta$$

$$B_0 \sum_{n=1}^{\infty} \frac{a}{n+1} \left(\frac{r}{a}\right)^{n+1} \left[\left(a_n + \frac{n+1}{n}\frac{a}{\rho} a_{n-1}\right) \sin\left((n+1)\theta\right)\right.$$

$$\left. - \left(b_n + \frac{n+1}{n}\frac{a}{\rho} b_{n-1}\right) \cos\left((n+1)\theta\right)\right]$$

$$+ \frac{aB_0}{\rho} \sum_{n=0}^{\infty} \frac{a}{n+1} \left(\frac{r}{a}\right)^{n+2} \left[a_n \sin\left(n\theta\right) - b_n \cos\left(n\theta\right)\right].$$

$$(4.105)$$

So we find that there is a "feed-up" of the $n-1^{\text{st}}$ harmonic coefficients to the nth harmonic in the first summation, but in the second summation there is an extra part which does not really look like a simple harmonic in that there is an extra power of r^2 for each $\sin(n\theta)$ and $\cos(n\theta)$ term.

4.8 Problem 4–8: $\cos\theta$ dipole magnet

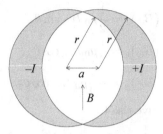

Fig. 4.6 The ideal cross section of a $\cos\theta$ style magnet. The central intersection of two identical circles will have a uniform magnetic field, if equal and opposite currents flow in the two outer lunes.

A magnet is constructed from conductors with a transverse cross section as shown in Fig. 4.6. Show that there is a uniform magnetic field in the region of intersection of the two circles.

Solution[71]:

Let us consider separately the fields created by each conductor as if the other conductor were not existing. Under this hypothesis, both circulations are represented by the broken line circles shown in Fig. 4.7; hence we can write

$$(B_\odot)_x = -B_\odot \sin\theta_\odot, \qquad (B_\odot)_y = B_\odot \cos\theta_\odot, \qquad (4.106)$$

and

$$(B_\oplus)_x = B_\oplus \sin\theta_\oplus, \qquad (B_\oplus)_x = -B_\oplus \cos\theta_\oplus. \qquad (4.107)$$

Moreover, recalling the discussion in Problem 4–1, we have

$$B_\odot = |\vec{B}_\odot| = \frac{1}{2}\mu_0 J r_\odot, \quad \text{and} \quad B_\oplus = |\vec{B}_\oplus| = \frac{1}{2}\mu_0 J r_\oplus, \qquad (4.108)$$

which transforms Eqs. (4.106 and 4.107) into

$$(B_\odot)_x = -\frac{1}{2}\mu_0 J r_\odot \sin\theta_\odot, \qquad (4.109)$$

$$(B_\oplus)_x = \frac{1}{2}\mu_0 J r_\oplus \sin\theta_\oplus, \qquad (4.110)$$

$$(B_\odot)_y = \frac{1}{2}\mu_0 J r_\odot \cos\theta_\odot, \qquad (4.111)$$

$$(B_\oplus)_y = -\frac{1}{2}\mu_0 J r_\oplus \cos\theta_\oplus, \qquad (4.112)$$

and which in their turn give rise to

$$B_x = (B_\odot)_x + (B_\oplus)_x = \frac{1}{2}\mu_0 J(-r_\odot \sin\theta_\odot + r_\oplus \sin\theta_\oplus) = 0, \qquad (4.113)$$

$$B_y = (B_\odot)_y + (B_\oplus)_y = \frac{1}{2}\mu_0 J(r_\odot \cos\theta_\odot - r_\oplus \cos\theta_\oplus) = -\frac{1}{2}\mu_0 J a, \qquad (4.114)$$

which is constant, as can be easily understood by giving a glance at Fig. 4.6. In fact, as far as Eq. (4.113) is concerned the dotted line demonstrates how the two projections coincide. With regard to Eq. (4.114), it is evident how the distance a is equal to the difference between the other two projections.

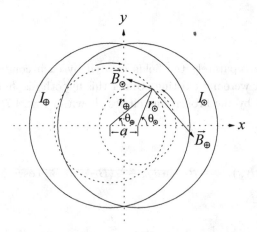

Fig. 4.7 Note that I_+ and I_- have be relabeled as $I_\odot$ (out of page) and $I_\oplus$ (into page), respectively to better emphasize the direction of current flow in each lune.

4.9   Problems 4–9: $\cos(n\theta)$ multipole magnet

Show that current distributed in a thin cylindrical shell with a strength $I(\theta) = I_0 \cos(n\theta)$, will produce a pure $2n$-multipole distribution inside the cylinder.

Solution:

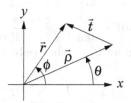

Fig. 4.8   A wire of current I flowing out $(+\hat{z})$ of the page is located at $\vec{\rho}$.

First consider the field from a single wire of current I located at $\vec{\rho}$ as illustrated in Fig. 4.8. The field at $\vec{r}$ will be

$$\vec{B}(r,\phi) = \frac{\mu_0 I}{2\pi t}(\hat{z} \times \hat{t}) = \frac{\mu_0 I}{2\pi} \frac{\hat{z} \times \vec{t}}{t^2}. \tag{4.115}$$

To get the direction of the field we have:

$$\vec{\rho} = \rho \cos\theta \, \hat{x} + \rho \sin\theta \, \hat{y}, \tag{4.116}$$

$$\vec{r} = r \cos\phi \, \hat{x} + r \sin\phi \, \hat{y}, \tag{4.117}$$

$$\vec{t} = (r\cos\phi - \rho\cos\theta)\hat{x} + (r\sin\phi - \rho\sin\theta)\hat{y}, \tag{4.118}$$

$$t = |\vec{r} - \vec{\rho}| = \sqrt{\rho^2 + r^2 - 2r\rho\cos(\phi - \theta)}, \tag{4.119}$$

$$\vec{B}(r,\phi) = \frac{\mu_0 I}{2\pi} \frac{-(r\sin\phi - \rho\sin\theta)\hat{x} + (r\cos\phi - \rho\cos\theta)\hat{y}}{r^2 + \rho^2 - 2r\rho\cos(\phi - \theta)}$$

$$= \frac{\mu_0 I}{2\pi\rho} \frac{(\sin\theta - \frac{y}{\rho})\hat{x} + (\frac{x}{\rho} - \cos\theta)\hat{y}}{1 + \frac{x^2+y^2}{\rho} - \frac{2x}{\rho}\cos\theta - \frac{2y}{\rho}\sin\theta}. \tag{4.120}$$

Recalling the multipole expansion Eq. (CM: 4.8)

$$B_x - iB_y = B_0 \sum_{n=0}^{\infty} (a_n - ib_n) \left(\frac{x + iy}{a}\right)^n, \tag{CM: 4.8}$$

we can express the field components along the horizontal axis as

$$B_x(x, y = 0) = B_0 \sum_{m=0}^{\infty} a_m \xi^m, \tag{4.121}$$

$$B_y(x, y = 0) = B_0 \sum_{m=0}^{\infty} b_m \xi^m, \tag{4.122}$$

where we have taken the reference radius $a = \rho$ and made the substitution $\xi = x/\rho$ to simplify the algebra. Then Eq. (4.120) with $y = 0$ becomes

$$B_x(x, 0) = \frac{\mu_0 I}{2\pi\rho} \frac{\sin\theta}{1 + \xi^2 - 2\xi\cos\theta}, \tag{4.123}$$

$$B_y(x, 0) = \frac{\mu_0 I}{2\pi\rho} \frac{\xi - \cos\theta}{1 + \xi^2 - 2\xi\cos\theta}. \tag{4.124}$$

The generating function for Chebyshev polynomials[7, 3] of the second kind is

$$(1 - 2\xi t + \xi^2)^{-1} = \sum_{j=0}^{\infty} \xi^j U_j(t), \tag{4.125}$$

and with $t = \cos\theta$, we have

$$U_j(\cos\theta) = \frac{\sin((j+1)\theta)}{\sin\theta}. \tag{4.126}$$

This gives

$$B_x(x, 0) = \frac{\mu_0 I}{2\pi\rho} \sum_{j=0}^{\infty} \xi^j \sin((j+1)\theta), \tag{4.127}$$

so comparing with Eq. (4.121) we find

$$B_0 a_m = \frac{\mu_0 I}{2\pi\rho} \sin((m+1)\theta). \tag{4.128}$$

Expanding the vertical component of the field:

$$
\begin{aligned}
B_y(x, 0) &= \frac{\mu_0 I}{\pi\rho} \frac{\xi - \cos\theta}{1 + \xi^2 - 2\xi\cos\theta} \\
&= \frac{\mu_0 I}{\pi\rho} (\xi - \cos\theta) \sum_{j=0}^{\infty} \xi^j U_j(\cos\theta) \\
&= \frac{\mu_0 I}{2\pi\rho} (\xi - \cos\theta) \sum_{j=0}^{\infty} \xi^j \frac{\sin((j+1)\theta)}{\sin\theta} \\
&= -\frac{\mu_0 I}{2\pi\rho} \left[\cos\theta + \sum_{j=1}^{\infty} \xi^j \left(\frac{\cos\theta\sin((j+1)\theta)}{\sin\theta} - \frac{\sin(j\theta)}{\sin\theta} \right) \right] \\
&= -\frac{\mu_0 I}{2\pi\rho} \sum_{j=0}^{\infty} \xi^j \cos((j+1)\theta),
\end{aligned}
\tag{4.129}
$$

which yields

$$B_0 b_m = -\frac{\mu_0 I}{2\pi\rho}\cos\left((m+1)\theta\right). \tag{4.130}$$

on comparison with Eq. (4.122).

The field for a current distribution $I(\theta) = I_0\cos(n\theta)$ on a thin shell may now be obtained by integrating θ over the shell:

$$B_x(x,0) = \frac{\mu_0 I_0}{2\pi\rho}\int_{-\pi}^{\pi}\frac{\cos(n\theta)\sin\theta}{1+\xi^2-2\xi\cos\theta}\,d\theta = 0, \tag{4.131}$$

since the integrand is an odd function of θ. Similarly for the B_y along the x-axis,

$$\begin{aligned}
B_y(x,0) &= \frac{\mu_0 I_0}{2\pi\rho}\int_{-\pi}^{\pi}\frac{\cos(n\theta)(\xi-\cos\theta)}{1+\xi^2-2\xi\cos\theta}\,d\theta \\
&= -\frac{\mu_0 I_0}{2\pi\rho}\sum_{j=0}^{\infty}\xi^j\int_{-\pi}^{\pi}\cos(n\theta)\cos\left((j+1)\theta\right)\,d\theta \\
&= -\frac{\mu_0 I_0}{2\rho}\xi^{n-1},
\end{aligned} \tag{4.132}$$

which we see is a pure $2n$-multipole field. (Recall Eq. (4.122).)

4.10 Problem 4–10: $\cos(n\theta)$ multipole magnet (Oops!)

Problem 4–10 is essentially a duplicate of 4–9 but has a funny normalization for the current, and it should have been deleted.

4.11   Problem 4–11: Deflectors: electric vs magnetic

A beam must be bent by an angle $\theta = 1$ mrad by an element of length $\ell = 0.5$ m. Compare the required fields needed for both an electrostatic (parallel plate) and a magnetic (dipole magnet) deflector for proton beams of kinetic energy 10 MeV, 100 MeV, 1 GeV, 10 GeV, and 100 GeV. What voltages would be required if the parallel plates were separated by 5 cm?

Solution:

Let us begin with the magnetic bending where the trajectories are arcs of circle of radius ρ. Therefore, recalling Eq. (CM: 1.25)

$$p = \beta\gamma mc = qB\rho, \qquad \text{(CM: 1.25)}$$

we have

$$\theta \simeq \sin\theta = \frac{\ell}{\rho} = \frac{qB\ell}{p}, \qquad (4.133)$$

or

$$B = \frac{\theta p}{q\ell}. \qquad (4.134)$$

Expressing the momentum p as a function of the kinetic energy W,

$$pc = \sqrt{(W + mc^2)^2 - (mc^2)^2} = W\sqrt{1 + 2\left(\frac{mc^2}{W}\right)}, \qquad (4.135)$$

and bearing in mind that $mc^2 = 0.938$ GeV for protons, we can obtain the results listed in Table 4.1.

Table 4.1   Magnetic Deflection

W [GeV]	p [Gev/c]	B [gauss]
0.01	0.137	9.14
0.10	0.445	29.65
1.00	1.696	113
10.00	10.90	727
100.00	100.93	6734

(NB: 1 gauss $= 10^{-4}$ tesla).

Considering now the electrostatic deflection, we have

$$\frac{d\vec{p}}{dt} = q\vec{E} \Rightarrow \begin{cases} \frac{dp_x}{dt} = qE, \\ \frac{dp_y}{dt} = 0, \\ \frac{dp_z}{dt} = 0; \end{cases} \Rightarrow \begin{cases} p_x = qEt \simeq \frac{qE\ell}{v}, \\ p_y = 0, \\ p_z \simeq p \quad \text{(paraxial approximation)}, \end{cases}$$

$$(4.136)$$

that altogether will yield

$$\theta \simeq \frac{p_x}{p} = \frac{qE\ell}{vp}, \tag{4.137}$$

from which it is possible to deduce

$$E = \theta \left[\frac{v}{c}\right] \left[\frac{pc/q}{\ell}\right] = \theta\beta E_{eq}. \tag{4.138}$$

In Table 4.2 we list the values of the fields E together with the voltages, $V = E\ell$, required for generating these fields.

Table 4.2 Electrostatic Deflection

W [GeV]	β	E [Volt/m]	V [Volt]
0.01	0.145	3.97×10^4	1.99×10^4
0.10	0.428	3.81×10^5	1.90×10^5
1.00	0.875	2.97×10^6	1.48×10^6
10.00	0.996	2.17×10^7	1.09×10^7
100.00	1.000	2.02×10^8	1.01×10^8

Let us proceed as follows:

- pick up the deflecting magnetic fields from the last column of Table 4.1;
- express them in tesla via simple division by 10000;
- write the velocities corresponding to the various kinetic energies multiplying the β's of the second column of Table 4.2 by the speed of light $c = 2.998 \times 10^8$ [m/s].

Hence, we shall be able to verify that the products vB equalize the values E. This is typical of the paraxial approximation, where the trajectories are very similar independently of the bending field. Moreover, this property is exploited for making spins precess while leaving the beam undeflected. In fact, choosing in the due way the directions of $\vec{B}$ and $\vec{E}$, the general expression of the Lorentz force may become null, i.e.

$$\frac{d\vec{p}}{dt} = q[\vec{E} - \vec{v} \times \vec{B}] = 0, \tag{4.139}$$

which means that $|\vec{E}|$ is equal to $|\vec{v}||\vec{B}|$. A comment deserves to be made. Electrostatic deflection implies a variation of the particle energy which, in the paraxial approximation, can be neglected. Nevertheless, in the chosen (z, x)-plane, an energy increment $\pm qE\delta x$ is generated by any lateral drift δx. No energy variation takes place in the example of the magnetic deflection, since the Lorentz force is always perpendicular to the motion direction; that is

$$dU = \vec{F} \cdot d\vec{\ell} = (\vec{v} \times \vec{B} \cdot \vec{v})dt = 0, \qquad (4.140)$$

since $d\vec{\ell} = \vec{v}dt$.

4.12 Problem 4–12: Dipole magnet design considerations

Consider a 0.5 m long window-frame dipole with the cross section shown below.

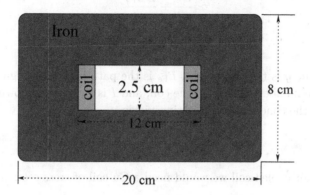

a) Estimate the number of ampere-turns needed to achieve a 0.6 T field in the gap. Assume that the iron is not saturated (i.e., $\mu_r \gtrsim 5000$).

b) Air-cooled copper coils can carry as much as 1.5 A/mm²; whereas water-cooled copper can carry almost 10 times as much (averaged over conductor and water channels). For the given dimensions of the magnet, would you recommend air-cooled or water-cooled coils. How wide would the gap between the coils be?

c) If the magnet is to be powered by a supply with a maximum current of 1000 A, how many turns should be in the coil.

d) What is the stored energy in the gap.

e) Assuming a constant field in the iron, estimate the additional energy stored in the iron yoke.

f) Estimate the inductance of the magnet.

Solution:

a) Recall from your basic electromagnetism courses that the component of magnetic induction $\vec{B}_\perp$ perpendicular to a boundary between two media (air and iron) is continuous across the boundary. We can approximate the magnitude of B around one of the legs of the coil as a constant with the field in the air gap perpendicular to the pole faces. Using Ampere's law to integrate around one of the legs of the coil, we have

$$NI = \oint \vec{H} \cdot d\vec{l} = \oint \frac{1}{\mu} \vec{B} \cdot d\vec{l}$$

$$\simeq \frac{Bg}{\mu_0} + \frac{BL_{Fe}}{\mu_0 \mu_r}$$

$$\simeq \frac{B}{\mu_0} \left(g + \frac{L_{Fe}}{\mu_r} \right) \tag{4.141}$$

where the g is the gap height, L_{Fe} is the path length through the iron, N is the number of turns in the coil, and I is the current flowing in one turn of the conductor. Evaluating yields

$$NI \simeq \frac{0.6 \text{ [T]} \times 0.025 \text{ [m]}}{4\pi \times 10^{-7} \text{ [Tm/A]}} = 11.94 \text{ kA}. \tag{4.142}$$

b) For an air cooled coil we would need a Cu coil of cross section

$$\frac{11940 \text{ [kA]}}{1.5 \text{ [A/mm}^2\text{]}} = 79.6 \text{ [cm}^2\text{]}. \tag{4.143}$$

Since the gap height is only 2.5 cm, this would require a width of 79.6/2.5=31.8 cm per leg, but the overall opening for the gap and coil is just 12 cm. A Cu air-cooled coil will not fit in this magnet. For a water-cooled coil, the leg width would only be 3.18 cm, and would leave an acceptable width of $12 - 2 \times 3.18 = 5.6$ cm in the gap.

c)

$$N = \frac{11.94 \text{ [kA]}}{1 \text{ [kA]}} = 11.94 \simeq 12 \text{ turns}, \tag{4.144}$$

after rounding upward. Note that this requires $I \simeq 11940/12 = 995$ A for full field.

d) With $g = 2.5$ cm, and a magnet length $l = 0.5$ m, the energy in the gap (excluding Cu conductors) is

$$U_{gap} = \frac{1}{2} \int \vec{B} \cdot \vec{H} \, dV = \frac{B^2 l w g}{2\mu_0}$$

$$\simeq \frac{(0.6 \text{ [T]})^2}{2 \times 4\pi \times 10^{-7} \text{ [Tm/A]}} \times 0.025 \text{ m} \times 0.056 \text{ m} \times 0.5 \text{ m}$$

$$\simeq 101 \text{ [J]}. \tag{4.145}$$

To estimate the energy stored in the conductors, let us try three possible approximations:

1. Assume a linear drop of field across the each conductor block from $x = -1.09$ to 1.09 cm as

$$B(x) = \frac{2 \times 0.6 \text{ [T]}}{0.0318 \text{ [m]}} x = 37.7[\text{T/m}]\, x, \qquad (4.146)$$

so

$$U_{\text{cond},1} = 2 \times lg \frac{(37.7 \text{ [T/m]})^2}{2\mu_0} \int_{-0.0109}^{0.0109} x^2\, dx = 6 \text{ J}. \qquad (4.147)$$

This is probably an under estimate, since we have ignored the horizontal field components in the conductors; the conductors are not infinitely high.

2. Assume the field in the conductors is the same as in the gap (clearly an over estimate):

$$U_{\text{cond},1} = \frac{lg}{2\mu_0} \times (0.6 \text{ T})^2 \times 2 \times 0.0318 \text{ [m]} = 114 \text{ J}. \qquad (4.148)$$

3. If we look at an isolated circular conductor of radius a, we would estimate the energy stored inside the conductor as

$$
\begin{aligned}
U_{\text{cond},3} &= \frac{1}{2\mu_0} \int_0^l \int_0^a \int_0^{2\pi} \left[\frac{\mu_0 \left(I \frac{\pi r^2}{\pi a^2} \right)}{2\pi r} \right]^2 r\, d\theta\, dr\, dz \\
&= \frac{\mu_0 l I^2}{4\pi a^4} \int_0^a r^3\, dr = \frac{\mu_0 I^2 l}{16\pi} \\
&= \frac{1}{2} \times \left[\frac{1}{2\mu_0} \left(\frac{\mu_0 I}{2\pi a} \right)^2 \pi a^2 l \right],
\end{aligned} \qquad (4.149)
$$

so we find that the energy in this case is half of what we would calculate by assuming that the field inside the conductor was constant field with the same value as at the surface of the conductor. Assuming this type of fall-off, we obtain

$$U_{\text{cond},3} = \frac{1}{2} U_{\text{cond},2} = 57 \text{ J}. \qquad (4.150)$$

This third estimate may be closer to reality than the first or second, so that we may estimate the total energy inside the gap plus conductors as

$$U_{\text{hole}} \approx U_{\text{gap}} + U_{\text{cond},3} \approx 150 \text{ J}. \qquad (4.151)$$

e) In the iron the stored energy is

$$U_{\text{Fe}} = \frac{1}{2} \int \vec{B} \cdot \vec{H}\, dV = \int \frac{B^2}{2\mu_r \mu_0}\, dV. \qquad (4.152)$$

For a constant magnitude of field in the iron, the energy density is

$$\frac{B^2}{2\mu_r\mu_0} \lesssim \frac{(0.6 \text{ [T]})^2}{2 \times 5000 \times 4\pi \times 10^{-7} \text{ [Tm/A]}} \simeq 28.65 \text{ [J/m}^3\text{]}, \qquad (4.153)$$

and the volume of iron is

$$0.5 \times (0.08 \times 0.2 - 0.025 \times 0.12) = 0.0065 \text{ [m}^3\text{]}. \qquad (4.154)$$

Multiplying these gives

$$U_{\text{Fe}} = 0.043 \text{ J}. \qquad (4.155)$$

This is negligible.

f) Recall that in terms of the inductance L and current I we may write the total stored energy as

$$U_{\text{tot}} = \frac{1}{2}LI^2. \qquad (4.156)$$

Solving for inductance gives

$$L = \frac{2U_{\text{tot}}}{I^2} = \frac{2 \times 150 \text{ [J]}}{(995 \text{ [A]})^2} \simeq 0.3 \text{ [mH]}. \qquad (4.157)$$

4.13 Problem 4–13: Pole-face profile of gradient dipole

An iron dominated dipole magnet has tilted pole faces. If we assume that the iron is not saturated, find a relation between the field index and gap height as a function of radial distance (x) from the design trajectory.

Solution:

Assume a C-magnet with a total number of turns N in the coils with at current I in one turn of the conductor. Writing Ampere's law in integral from, we have

$$\oint \vec{H} \cdot d\vec{l} = NI. \tag{4.158}$$

This leads to Eq. (4.141) of the previous problem, so we have

$$B_y(x) \simeq \frac{\mu_0 NI}{g(x) + L_{\text{Fe}}(x)/\mu_r}. \tag{4.159}$$

The L_{Fe}/μ_r can be thought of as a correction to the gap height allowing for a noninfinite permeability of the iron.

Assuming that the relative permeability of the iron $\mu_r \gg 1$, we will ignore the path length through the iron, so

$$B_y(x,) \simeq \frac{\mu_0 NI}{g(x)}, \tag{4.160}$$

for a gap height of $g(x)$. Recall that the field index was as defined in Eq. (CM: 2.13) as

$$n = -\frac{\rho}{B_0} \left(\frac{\partial B_y}{\partial x} \right)_{x=0}, \tag{CM: 2.13}$$

we we have

$$n = -\frac{\rho}{B_0} \left(-\frac{\mu_0 NI g_0'}{g^2} \right) = \rho \frac{g'}{g} = \rho \frac{d}{dx}(\ln g). \tag{4.161}$$

If we were to keep the L_{Fe}/μ_r correction, then we would have

$$\frac{n}{\rho} = \frac{d}{dx}[\ln(g + L_{\text{Fe}}/\mu_r)]. \tag{4.162}$$

4.14   Problem 4–14: Helical dipole field

Consider a helical dipole magnet twisted through a full $360°$ in a length λ with approximate transverse field components given by

$$B_x \simeq -B_0 \left\{ \left[1 + \frac{k^2}{8}(3x^2 + y^2) \right] \sin kz - \frac{k^2}{4}xy \cos kz \right\}, \qquad (4.163)$$

$$B_y \simeq B_0 \left\{ \left[1 + \frac{k^2}{8}(x^2 + 3y^2) \right] \cos kz - \frac{k^2}{4}xy \sin kz \right\}, \qquad (4.164)$$

$$B_z \simeq -B_0 k \left[1 + \frac{k^2}{8}(x^2 + y^2) \right] (x \cos kz + y \sin kz), \qquad (4.165)$$

where the pitch is $k = \frac{2\pi}{\lambda}$.

a) Show that this parameterization satisfies Maxwell's equations to second order in the transverse dimensions.

b) Ignoring the transverse dependence of the field

$$(B_x \simeq -B_0 \sin kz, \quad B_y \simeq B_0 \cos kz, \quad \text{and} \quad B_z \simeq 0) \qquad (4.166)$$

find an approximation for the trajectory of a particle through the magnet in the paraxial approximation.

Solution:

a) For Gauss' law $\nabla \cdot \vec{B} = 0$, we have the terms

$$\frac{\partial B_x}{\partial x} = -B_0 k^2 \left[\frac{3x}{4} \sin kz - \frac{y}{4} \cos kz \right], \qquad (4.167)$$

$$\frac{\partial B_y}{\partial y} = B_0 k^2 \left[\frac{3y}{4} \cos kz - \frac{x}{4} \sin kz \right], \qquad (4.168)$$

$$\frac{\partial B_z}{\partial z} = -B_0 k^2 \left[1 + \frac{k^2}{8}(x^2 + y^2) \right] (-x \sin kz + y \cos kz), \qquad (4.169)$$

which add up to give

$$\nabla \cdot \vec{B} = B_0 k^2 \left[\left(-\frac{3}{4} - \frac{1}{4} + 1 \right) x \sin kz + \left(\frac{1}{4} + \frac{3}{4} - 1 \right) y \cos kz \right]$$
$$- B_0 \frac{k^3}{8}(x^2 + y^2)(x \cos kz + y \sin kz)$$
$$= -B_0 \frac{k^3}{8}(x^2 + y^2)(x \cos kz + y \sin kz) \simeq 0 + \mathcal{O}(3).$$

$$(4.170)$$

For Ampere's law $\nabla \times \vec{B} = 0$,

$$\frac{\partial B_y}{\partial x} - \frac{\partial B_x}{\partial y} = B_0 k^2 \left[\frac{x}{4} \cos kz - \frac{y}{4} \sin kz \right]$$
$$+ B_0 k^2 \left[\frac{y}{4} \sin kz - \frac{x}{4} \cos kz \right] = 0, \qquad (4.171)$$

$$\frac{\partial B_z}{\partial y} - \frac{\partial B_y}{\partial z} = -B_0 k \left\{ \left[1 + \frac{k^2(x^2 + 3y^2)}{8} \right] \sin kz + \frac{k^2 xy}{4} \cos kz \right\}$$
$$- B_0 k \left\{ - \left[1 + \frac{k^2}{8}(x^2 + 3y^2) \right] \sin kz - \frac{k^2 xy}{4} \cos kz \right\}$$
$$= 0, \qquad (4.172)$$

$$\frac{\partial B_x}{\partial z} - \frac{\partial B_z}{\partial x} = -B_0 k \left\{ \left[1 + \frac{k^2(3x^2 + y^2)}{8} \right] \cos kz + \frac{k^3 xy}{8} \sin kz \right\}$$
$$+ B_0 k \left\{ \left[1 + \frac{k^2(3x^2 + y^2)}{8} \right] \cos kz + \frac{k^2 xy}{8} \sin kz \right\}$$
$$= 0. \qquad (4.173)$$

b) From the Lorentz force, we have

$$\frac{d}{dt} \begin{pmatrix} p_x \hat{x} \\ p_y \hat{y} \end{pmatrix} \simeq q \frac{dz}{dt} \hat{z} \times \begin{pmatrix} \hat{y} \, B_0 \cos kz \\ -\hat{x} \, B_0 \sin kz \end{pmatrix}. \qquad (4.174)$$

Rearranging and dividing by the design momentum p_0 gives the simple differential equations for the transverse positions:

$$\begin{pmatrix} x'' \\ y'' \end{pmatrix} \simeq \frac{1}{p_0} \frac{d}{dz} \begin{pmatrix} p_x \\ p_y \end{pmatrix} \simeq -\frac{qB_0}{p_0} \begin{pmatrix} \cos kz \\ \sin kz. \end{pmatrix} \qquad (4.175)$$

Integrating once yields the slopes

$$\begin{pmatrix} x' \\ y' \end{pmatrix} = \frac{qB_0}{p_0 k} \begin{pmatrix} \sin kz \\ 1 - \cos kz \end{pmatrix} + \begin{pmatrix} x'_0 \\ y'_0 \end{pmatrix}, \qquad (4.176)$$

where x'_0 and y'_0 are the respective slopes at $z = 0$. After another integration, we obtain the positions

$$\begin{pmatrix} x \\ y \end{pmatrix} = \begin{pmatrix} x'_0 \\ y'_0 \end{pmatrix} z - \frac{qB_0}{p_0 k^2} \begin{pmatrix} -\cos kz \\ \sin kz - z \end{pmatrix} + \begin{pmatrix} a \\ b \end{pmatrix} \qquad (4.177)$$

where a and b must be determined from the initial positions x_0 and y_0. Solving for these new integration constants, we find

$$\begin{pmatrix} a \\ b \end{pmatrix} = \begin{pmatrix} x_0 - \frac{qB_0}{p_0 k^2} \\ y_0 \end{pmatrix}, \qquad (4.178)$$

or putting it all together:

$$\begin{pmatrix} x \\ y \end{pmatrix} = \begin{pmatrix} x_0 \\ y_0 \end{pmatrix} + \begin{pmatrix} x'_0 \\ y'_0 \end{pmatrix} z + \frac{qB_0}{p_0 k^2} \begin{pmatrix} \cos kz - 1 \\ z - \sin kz \end{pmatrix}. \qquad (4.179)$$

4.15   Problem 4–15: Multipole vector potential

a) Using Eqs. (CM:4.6 & CM:4.7)

$$B_y = B_0 \sum_{n=0}^{\infty} \left(\frac{r}{a}\right)^n (b_n \cos n\theta - a_n \sin n\theta), \qquad \text{(CM:4.6)}$$

$$B_x = B_0 \sum_{n=0}^{\infty} \left(\frac{r}{a}\right)^n (a_n \cos n\theta + b_n \sin n\theta), \qquad \text{(CM:4.7)}$$

show that the field in the polar coordinate representation is

$$B_r = B_0 \sum_{n=0}^{\infty} \left(\frac{r}{a}\right)^n [a_n \cos((n+1)\theta) + b_n \sin((n+1)\theta)], \qquad (4.180)$$

$$B_\theta = B_0 \sum_{n=0}^{\infty} \left(\frac{r}{a}\right)^n [b_n \cos((n+1)\theta) - a_n \sin((n+1)\theta)]. \qquad (4.181)$$

b) Show that the vector potential [sic]

$$A_r = 0, \qquad (4.182)$$

$$A_\theta = 0, \qquad (4.183)$$

$$A_z = B_0 \sum_{n=0}^{\infty} \frac{a}{n+1} \left(\frac{r}{a}\right)^{n+1} [a_n \sin((n+1)\theta) - b_n \sin((n+1)\theta)], \qquad (4.184)$$

leads to the components given in part (a).

Solution:

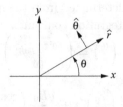

Fig. 4.9   Orientation of rotation of axes from (x, y) to (r, θ) coordinates.

a) Conversion of unit vectors from polar to Cartesian coordinates is given by (see Fig. 4.9)

$$\begin{pmatrix} \hat{x} \\ \hat{y} \end{pmatrix} = \begin{pmatrix} \cos\theta & -\sin\theta \\ \sin\theta & \cos\theta \end{pmatrix} \begin{pmatrix} \hat{r} \\ \hat{\theta} \end{pmatrix}. \tag{4.185}$$

The field in vector components is then

$$\begin{aligned} \vec{B} &= B_x\hat{x} + B_y\hat{y} \\ &= (B_x\cos\theta + B_y\sin\theta)\hat{r} + (-B_x\sin\theta + B_y\cos\theta)\hat{\theta}, \end{aligned} \tag{4.186}$$

so the polar components of the field are

$$\begin{aligned} B_r &= B_0 \sum_{n=0}^{\infty} \left(\frac{r}{a}\right)^n [\cos\theta(a_n\cos n\theta + b_n\sin n\theta) \\ &\qquad\qquad + \sin\theta(b_n\cos n\theta - a_n\sin n\theta)] \\ &= B_0 \sum_{n=0}^{\infty} \left(\frac{r}{a}\right)^n [a_n\cos((n+1)\theta) + b_n\sin((n+1)\theta)], \end{aligned} \tag{4.187}$$

$$\begin{aligned} B_\theta &= B_0 \sum_{n=0}^{\infty} \left(\frac{r}{a}\right)^n [-\sin\theta(a_n\cos n\theta + b_n\sin n\theta) \\ &\qquad\qquad + \cos\theta(b_n\cos n\theta - a_n\sin n\theta)] \\ &= B_0 \sum_{n=0}^{\infty} \left(\frac{r}{a}\right)^n [-a_n\sin((n+1)\theta) + b_n\cos((n+1)\theta)], \end{aligned} \tag{4.188}$$

which agree with Eqs. (4.180 and 4.181).

b) Oops! Well that answer is obviously wrong: the trigonometric coefficients of a_n and b_n cannot both be sine functions. When presented with a typo like this, we should try to find the correct answer.

With $A_r = A_\theta = 0$, we have

$$\begin{aligned} \vec{B} &= \nabla \times \vec{A} \\ &= \frac{1}{r}\frac{\partial A_z}{\partial\theta}\hat{r} - \frac{\partial A_z}{\partial r}\hat{\theta}. \end{aligned} \tag{4.189}$$

To find a viable expression for A_z we can integrate the expressions for B_r and B_θ:

$$A_z(r,\theta) = \int rB_r\,d\theta + C_1(r), \tag{4.190}$$

$$A_z(r,\theta) = -\int B_\theta\,dr + C_2(\theta), \tag{4.191}$$

where $C_1(r)$ is independent of θ, and $C_2(\theta)$ is independent of r. Using Eq. (4.190) with Eq. (4.180) gives

$$A_z = B_0 \sum_{n=0}^{\infty} a \left(\frac{r}{a}\right)^{n+1} \int \left[a_n \cos\left((n+1)\theta\right) + b_n \sin\left((n+1)\theta\right)\right] d\theta$$

$$= B_0 \sum_{n=0}^{\infty} \frac{a}{n+1} \left(\frac{r}{a}\right)^{n+1} \left[a_n \sin\left((n+1)\theta\right) - b_n \cos\left((n+1)\theta\right)\right]$$

$$+ C_1(r).$$

$$(4.192)$$

Similarly substituting Eq. (4.181) into Eq. (4.191) produces

$$A_z = -B_0 \sum_{n=0}^{\infty} \int \left(\frac{r}{a}\right)^{n} dr \left[-a_n \sin\left((n+1)\theta\right) + b_n \cos\left((n+1)\theta\right)\right]$$

$$= B_0 \sum_{n=0}^{\infty} \frac{a}{n+1} \left(\frac{r}{a}\right)^{n+1} \left[a_n \sin\left((n+1)\theta\right) - b_n \cos\left((n+1)\theta\right)\right]$$

$$+ C_2(\theta).$$

$$(4.193)$$

A comparison of Eqs. (4.192 and 4.193) reveals that we have equality if we simply set $C_1(r) = C_2(\theta) = 0$.

4.16 Problem 4–16: Symplecticity and solenoid fringes

Verify that $\mathbf{M}_{\text{body}}$ is not symplectic, but that $\mathbf{M}_{\text{sol}}$ is symplectic.

Solution:

Note that both $\mathbf{M}_{\text{body}}$ and $\mathbf{M}_{\text{sol}}$ have the similar 2×2 block forms:

$$\mathbf{M}_{\text{body}} = \begin{pmatrix} \mathbf{A} & \mathbf{B} \\ -\mathbf{B} & \mathbf{A} \end{pmatrix}, \quad \text{with} \tag{4.194}$$

$$\mathbf{A} = \begin{pmatrix} 1 & \frac{\sin kl}{k} \\ 0 & \cos kl \end{pmatrix}, \quad \text{and} \quad \mathbf{B} = \begin{pmatrix} 0 & \frac{1-\cos kl}{k} \\ 0 & \sin kl \end{pmatrix}, \tag{4.195}$$

$$\mathbf{M}_{\text{sol}} = \begin{pmatrix} \mathbf{C} & \mathbf{D} \\ -\mathbf{D} & \mathbf{C} \end{pmatrix}, \quad \text{with} \tag{4.196}$$

$$\mathbf{C} = \begin{pmatrix} \frac{1+\cos kl}{2} & \frac{\sin kl}{k} \\ -\frac{k\sin kl}{4} & \frac{1+\cos kl}{2} \end{pmatrix}, \quad \text{and} \quad \mathbf{D} = \begin{pmatrix} \frac{\sin kl}{2} & \frac{1-\cos kl}{k} \\ -k\frac{1-\cos kl}{4} & \frac{\sin kl}{2} \end{pmatrix}. \tag{4.197}$$

Applying the symplectic condition for this 2×2-block form, we would like to test if

$$\begin{pmatrix} \mathbf{S} & 0 \\ 0 & \mathbf{S} \end{pmatrix} \overset{?}{=} \begin{pmatrix} \mathbf{A}^{\mathsf{T}} & -\mathbf{B}^{\mathsf{T}} \\ \mathbf{B}^{\mathsf{T}} & \mathbf{A}^{\mathsf{T}} \end{pmatrix} \begin{pmatrix} \mathbf{S} & 0 \\ 0 & \mathbf{S} \end{pmatrix} \begin{pmatrix} \mathbf{A} & \mathbf{B} \\ -\mathbf{B} & \mathbf{A} \end{pmatrix}$$
$$= \begin{pmatrix} \mathbf{A}^{\mathsf{T}}\mathbf{S}\mathbf{A} + \mathbf{B}^{\mathsf{T}}\mathbf{S}\mathbf{B} & \mathbf{A}^{\mathsf{T}}\mathbf{S}\mathbf{B} - \mathbf{B}^{\mathsf{T}}\mathbf{S}\mathbf{A} \\ \mathbf{B}^{\mathsf{T}}\mathbf{S}\mathbf{A} - \mathbf{A}^{\mathsf{T}}\mathbf{S}\mathbf{B} & \mathbf{A}^{\mathsf{T}}\mathbf{S}\mathbf{A} + \mathbf{B}^{\mathsf{T}}\mathbf{S}\mathbf{B} \end{pmatrix} \tag{4.198}$$

with

$$\mathbf{S} = \begin{pmatrix} 0 & 1 \\ -1 & 0 \end{pmatrix}. \tag{4.199}$$

Notice that the two diagonal blocks in the last line of Eq. (4.198) are identical and the two off-diagonal blocks only differ by a sign, so for symplecticity, we must show

$$\mathbf{A}^{\mathsf{T}}\mathbf{S}\mathbf{A} + \mathbf{B}^{\mathsf{T}}\mathbf{S}\mathbf{B} = \mathbf{S}, \quad \text{and} \tag{4.200}$$
$$\mathbf{A}^{\mathsf{T}}\mathbf{S}\mathbf{B} - \mathbf{B}^{\mathsf{T}}\mathbf{S}\mathbf{A} = \mathbf{0}. \tag{4.201}$$

Applying the condition (4.200) to $\mathbf{M}_{\text{body}}$ we have

$$\mathbf{A}^{\mathsf{T}}\mathbf{S}\mathbf{A} = \begin{pmatrix} 1 & 0 \\ \frac{\sin kl}{k} & \cos kl \end{pmatrix} \begin{pmatrix} 0 & 1 \\ -1 & 0 \end{pmatrix} \begin{pmatrix} 1 & \frac{\sin kl}{k} \\ 0 & \cos kl \end{pmatrix}$$
$$= \begin{pmatrix} 1 & 0 \\ \frac{\sin kl}{k} & \cos kl \end{pmatrix} \begin{pmatrix} 0 & \cos kl \\ -1 & -\frac{\sin kl}{k} \end{pmatrix} = \begin{pmatrix} 0 & \cos kl \\ -\cos kl & 0 \end{pmatrix}, \tag{4.202}$$

and

$$\mathbf{B}^{\mathrm{T}}\mathbf{S}\mathbf{B} = \begin{pmatrix} 0 & 0 \\ \frac{1-\cos kl}{k} & \sin kl \end{pmatrix} \begin{pmatrix} 0 & 1 \\ -1 & 0 \end{pmatrix} \begin{pmatrix} 0 & \frac{1-\cos kl}{k} \\ 0 & \sin kl \end{pmatrix}$$

$$= \begin{pmatrix} 0 & 0 \\ \frac{1-\cos kl}{k} & \sin kl \end{pmatrix} \begin{pmatrix} 0 & \sin kl \\ 0 & -\frac{1-\cos kl}{k} \end{pmatrix}$$

$$= \begin{pmatrix} 0 & 0 \\ 0 & 0 \end{pmatrix}. \tag{4.203}$$

Adding these last two equations we find that

$$\mathbf{A}^{\mathrm{T}}\mathbf{S}\mathbf{A} + \mathbf{B}^{\mathrm{T}}\mathbf{S}\mathbf{B} = \begin{pmatrix} 0 & \cos kl \\ -\cos kl & 0 \end{pmatrix} \neq \mathbf{S}, \tag{4.204}$$

so $\mathbf{M}_{\text{body}}$ cannot be a symplectic matrix.

To check the symplecticity of $\mathbf{M}_{\text{sol}}$ we must evaluate:

$$\mathbf{C}^{\mathrm{T}}\mathbf{S}\mathbf{C} = \begin{pmatrix} \frac{1+\cos kl}{2} & -\frac{k\sin kl}{4} \\ \frac{\sin kl}{k} & \frac{1+\cos kl}{2} \end{pmatrix} \cdot \begin{pmatrix} 0 & 1 \\ -1 & 0 \end{pmatrix} \begin{pmatrix} \frac{1+\cos kl}{2} & \frac{\sin kl}{k} \\ -\frac{k\sin kl}{4} & \frac{1+\cos kl}{2} \end{pmatrix}$$

$$= \begin{pmatrix} \frac{1+\cos kl}{2} & -\frac{k\sin kl}{4} \\ \frac{\sin kl}{k} & \frac{1+\cos kl}{2} \end{pmatrix} \begin{pmatrix} -\frac{k\sin kl}{4} & \frac{1+\cos kl}{2} \\ -\frac{1+\cos kl}{2} & -\frac{\sin kl}{k} \end{pmatrix}$$

$$= \begin{pmatrix} 0 & \frac{1+2\cos kl+\cos^2 kl+\sin^2 kl}{4} \\ -\frac{\sin^2 kl+1+2\cos kl+\cos^2 kl}{4} & 0 \end{pmatrix}$$

$$= \begin{pmatrix} 0 & \frac{1+\cos kl}{2} \\ -\frac{1+\cos kl}{2} & 0 \end{pmatrix}, \tag{4.205}$$

$$\mathbf{D}^{\mathrm{T}}\mathbf{S}\mathbf{D} = \begin{pmatrix} \frac{\sin kl}{2} & -k\frac{1-\cos kl}{4} \\ \frac{1-\cos kl}{k} & \frac{\sin kl}{2} \end{pmatrix} \begin{pmatrix} 0 & 1 \\ -1 & 0 \end{pmatrix} \begin{pmatrix} \frac{\sin kl}{2} & \frac{1-\cos kl}{k} \\ -k\frac{1-\cos kl}{4} & \frac{\sin kl}{2} \end{pmatrix}$$

$$= \begin{pmatrix} \frac{\sin kl}{2} & -k\frac{1-\cos kl}{4} \\ \frac{1-\cos kl}{k} & \frac{\sin kl}{2} \end{pmatrix} \begin{pmatrix} -k\frac{1-\cos kl}{4} & \frac{\sin kl}{2} \\ -\frac{\sin kl}{2} & -\frac{1-\cos kl}{k} \end{pmatrix}$$

$$= \begin{pmatrix} 0 & \frac{\sin^2 kl+1-2\cos kl+\cos^2 kl}{4} \\ -\frac{1-2\cos kl+\cos^2 kl+\sin^2 kl}{4} & 0 \end{pmatrix}$$

$$= \begin{pmatrix} 0 & \frac{1-\cos kl}{2} \\ -\frac{1-\cos kl}{2} & 0 \end{pmatrix}, \tag{4.206}$$

and adding does indeed give $\mathbf{C}^{\mathrm{T}}\mathbf{S}\mathbf{C} + \mathbf{D}^{\mathrm{T}}\mathbf{S}\mathbf{D} = \mathbf{S}$. Next we must satisfy the second condition $\mathbf{C}^T\mathbf{SD} - \mathbf{D}^T\mathbf{SC} = 0$:

$$\mathbf{C}^T\mathbf{SD} = \begin{pmatrix} \frac{1+\cos kl}{2} & -\frac{k\sin kl}{4} \\ \frac{\sin kl}{k} & \frac{1+\cos kl}{2} \end{pmatrix} \begin{pmatrix} 0 & 1 \\ -1 & 0 \end{pmatrix} \begin{pmatrix} \frac{\sin kl}{2} & \frac{1-\cos kl}{k} \\ -k\frac{1-\cos kl}{4} & \frac{\sin kl}{2} \end{pmatrix}$$

$$= \begin{pmatrix} \frac{1+\cos kl}{2} & -\frac{k\sin kl}{4} \\ \frac{\sin kl}{k} & \frac{1+\cos kl}{2} \end{pmatrix} \begin{pmatrix} -k\frac{1-\cos kl}{4} & \frac{\sin kl}{2} \\ -\frac{\sin kl}{2} & -\frac{1-\cos kl}{k} \end{pmatrix}$$

$$= \begin{pmatrix} -k\frac{1-\cos^2 kl-\sin^2 kl}{8} & \frac{\sin kl(1+\cos kl)+\sin kl(1-\cos kl)}{4} \\ -\frac{\sin kl(1-\cos kl)+\sin kl(1+\cos kl)}{4} & \frac{\sin^2 kl+\cos^2 kl-1}{8} \end{pmatrix}$$

$$= \begin{pmatrix} 0 & \frac{\sin kl}{2} \\ -\frac{\sin kl}{2} & 0 \end{pmatrix}, \tag{4.207}$$

but everything is copacetic, since

$$\mathbf{C}^T\mathbf{SD} - \mathbf{D}^T\mathbf{SC} = \mathbf{C}^T\mathbf{SD} - (\mathbf{C}^T\mathbf{S}^T\mathbf{D})^T$$
$$= \mathbf{C}^T\mathbf{SD} + (\mathbf{C}^T\mathbf{SD})^T = 0. \tag{4.208}$$

Comments:

It is worth noting an important difference between Liouville's theorem and the symplectic condition. In his 1838 paper, Liouville[44, 51] did not refer to Hamilton nor canonical variables. Liouville's paper dealt with solutions to a system of first order differential equations of n variables:

$$\frac{d\vec{x}}{dt} = \vec{P}(\vec{x},t), \tag{4.209}$$

with solutions

$$\vec{x} = \vec{x}(\vec{a},t). \tag{4.210}$$

In somewhat archaic terms, he showed that the Jacobian determinant of the transformation in Eq. (4.210) from $\vec{a}$ to $\vec{x}$:

$$u(t) = \begin{vmatrix} \frac{\partial x_1}{\partial a_1} & \frac{\partial x_2}{\partial a_1} & \cdots & \frac{\partial x_n}{\partial a_1} \\ \frac{\partial x_1}{\partial a_2} & \frac{\partial x_2}{\partial a_2} & \cdots & \frac{\partial x_n}{\partial a_2} \\ \vdots & \vdots & \ddots & \vdots \\ \frac{\partial x_1}{\partial a_n} & \frac{\partial x_2}{\partial a_n} & \cdots & \frac{\partial x_n}{\partial a_n} \end{vmatrix}, \tag{4.211}$$

satisfies the differential equation

$$\frac{du}{dt} = (\nabla \cdot \vec{P})\,u. \tag{4.212}$$

If the trace $(\nabla \cdot \vec{P})$ of the matrix in Eq. (4.211) is zero, then the determinant u remains constant. In fact at the end of his paper, Liouville gave a formula for the growth rate of the volume element if the trace were not zero. This extra result is in effect a more general form of the Robinson sum rule[57].

It was Jacobi[35] who noticed the connection of Liouville's paper with Hamilton's equations. The symplectic requirement $\mathbf{M}^T\mathbf{S}\mathbf{M} = \mathbf{S}$ for the transport matrix $\mathbf{M}$ is more stringent than Liouville's theorem. For symplecticity, we must use canonical variables in accordance with the Hamiltonian, but for Liouville's theorem a Hamiltonian is not required. So long as the trace $\nabla \cdot \vec{P}$ is zero, the volume element will be preserved.

4.17   Problem 4–17: Solenoid rotational decomposition

Verify Eq. (CM: 4.78):

$$\mathbf{M}_{\text{sol}} = \mathbf{R}\mathbf{M}_f = \mathbf{M}_f\mathbf{R}. \qquad \text{(CM : 4.78)}$$

Solution:

Let us rewrite Eqs. (CM: 4.79), (CM: 4.80) and (CM: 4.81):

$$\mathbf{R} = \begin{pmatrix} \mathbf{I} \cos \frac{kl}{2} & \mathbf{I} \sin \frac{kl}{2} \\ -\mathbf{I} \sin \frac{kl}{2} & \mathbf{I} \cos \frac{kl}{2} \end{pmatrix} \qquad \text{(CM: 4.79)}$$

$$= \begin{pmatrix} \mathbf{I}c & \mathbf{I}s \\ -\mathbf{I}s & \mathbf{I}c \end{pmatrix}, \qquad (4.213)$$

$$\mathbf{M}_f = \begin{pmatrix} \mathbf{F} & \mathbf{0} \\ \mathbf{0} & \mathbf{F} \end{pmatrix}, \qquad \text{(CM: 4.80)}$$

$$\mathbf{F} = \begin{pmatrix} \cos \frac{kl}{2} & \frac{2}{k} \sin \frac{kl}{2} \\ -\frac{k}{2} \sin \frac{kl}{2} & \cos \frac{kl}{2} \end{pmatrix} \qquad \text{(CM: 4.81)}$$

$$= \begin{pmatrix} c & \frac{k}{2}s \\ -\frac{2}{k}s & c \end{pmatrix}, \qquad (4.214)$$

where

$$\mathbf{I} = \begin{pmatrix} 1 & 0 \\ 0 & 1 \end{pmatrix} \quad \text{is the } 2\times 2 \text{ identity matrix, and} \qquad (4.215)$$

$$\mathbf{0} = \begin{pmatrix} 0 & 0 \\ 0 & 0 \end{pmatrix} \quad \text{is the } 2\times 2 \text{ null matrix.} \qquad (4.216)$$

First of all let us verify the second equivalence of Eq. (CM: 4.78):

$$\mathbf{R}\mathbf{M}_f = \begin{pmatrix} \mathbf{I}c & \mathbf{I}s \\ -\mathbf{I}s & \mathbf{I}c \end{pmatrix} \begin{pmatrix} \mathbf{F} & \mathbf{0} \\ \mathbf{0} & \mathbf{F} \end{pmatrix} = \begin{pmatrix} c\mathbf{IF} & s\mathbf{IF} \\ -s\,\mathbf{IF} & c\mathbf{IF} \end{pmatrix} = \begin{pmatrix} c\mathbf{F} & s\mathbf{F} \\ -s\mathbf{F} & c\mathbf{F} \end{pmatrix}, \qquad (4.217)$$

$$\mathbf{M}_f\mathbf{R} = \begin{pmatrix} \mathbf{F} & \mathbf{0} \\ \mathbf{0} & \mathbf{F} \end{pmatrix} \begin{pmatrix} \mathbf{I}c & \mathbf{I}s \\ -\mathbf{I}s & \mathbf{I}c \end{pmatrix} = \begin{pmatrix} c\mathbf{FI} & s\mathbf{FI} \\ -s\mathbf{FI} & c\mathbf{FI} \end{pmatrix} = \begin{pmatrix} c\mathbf{F} & s\mathbf{F} \\ -s\mathbf{F} & c\mathbf{F} \end{pmatrix}. \qquad (4.218)$$

Since we have:

$$c\mathbf{F} = \begin{pmatrix} c^2 & \frac{k}{2}sc \\ -\frac{2}{k}sc & c^2 \end{pmatrix}, \quad \text{and} \quad s\mathbf{F} = \begin{pmatrix} sc & \frac{k}{2}s^2 \\ -\frac{2}{k}s^2 & sc \end{pmatrix}, \qquad (4.219)$$

we can write

$$\mathbf{M}_{\text{sol}} = \begin{pmatrix} c^2 & \frac{2}{k}sc & sc & \frac{2}{k}s^2 \\ -\frac{k}{2}sc & c^2 & -\frac{k}{2}s^2 & sc \\ sc & \frac{2}{k}s^2 & c^2 & \frac{2}{k}sc \\ -\frac{k}{2}s^2 & sc & -\frac{k}{2}sc & c^2 \end{pmatrix}. \tag{4.220}$$

This coincides with the matrix $\mathbf{M}_{\text{sol}}$, and which appears in Eqs. (CM: 4.77 and CM: 4.78), if we consider the following identities:

$$c^2 = \cos^2 \frac{kl}{2} = \frac{1 + \cos kl}{2}; \tag{4.221}$$

$$sc = \sin \frac{kl}{2} \cos \frac{kl}{2} = \frac{\sin kl}{2}; \tag{4.222}$$

$$\frac{2}{k}sc = \frac{\sin kl}{k}; \tag{4.223}$$

$$\frac{k}{2}sc = \frac{\sin kl}{2}; \tag{4.224}$$

$$\frac{2}{k}s^2 = \frac{2}{k}\frac{1 - \cos kl}{2} = \frac{1 - \cos kl}{k}; \tag{4.225}$$

$$\frac{k}{2}s^2 = \frac{k}{2}\frac{1 - \cos kl}{2} = k\frac{1 - \cos kl}{4}. \tag{4.226}$$

4.18   Problem 4–18: Principal planes of a thick quadrupole

Show that the focusing matrix $\mathbf{F}$ in Eq. (CM: 4:81)

$$\mathbf{F} = \begin{pmatrix} \cos\frac{kl}{2} & \frac{2}{k}\sin\frac{kl}{2} \\ -\frac{k}{2}\sin\frac{kl}{2} & \cos\frac{kl}{2} \end{pmatrix},$$ (CM : 4.81)

may be written as the product of a drift followed by a thin lens and then a second drift of the same length as the first.

Solution[64, 36]:

A thick lens quadrupole can be modelled as a thin lens surrounded by two drifts. First consider a parallel ray entering the quadrupole as shown in Fig. 4.10. In a focusing quadrupole, the effective focal length f will look as though there was a thin kick a distance h upstream of the middle of the magnet.

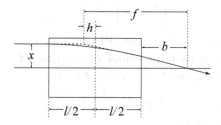

Fig. 4.10   Thick lens focusing quadrupole with a principal plane located at $l/2 - h$.

Write Eq. (CM: 4.81) in the form:

$$\mathbf{F} = \begin{pmatrix} \cos\phi & \frac{2}{k}\sin\phi \\ -\frac{k}{2}\sin\phi & \cos\phi \end{pmatrix},$$ (4.227)

with $\phi = kl/2$. (Note that this looks the same as a thick-lens quadruple matrix of Eq.(CM:32), but with $\sqrt{k}$ replaced by $k/2$.) A parallel ray with transverse offset x will have the position

$$\begin{pmatrix} x_e \\ x'_e \end{pmatrix} = \begin{pmatrix} \cos\phi & \frac{2}{k}\sin\phi \\ -\frac{k}{2}\sin\phi & \cos\phi \end{pmatrix} \begin{pmatrix} x \\ 0 \end{pmatrix} = x \begin{pmatrix} \cos\phi \\ -\frac{k}{2}\sin\phi \end{pmatrix},$$ (4.228)

at the end of the magnet. Continuing on to the focal point gives

$$\begin{pmatrix} x_f \\ x'_f \end{pmatrix} = \begin{pmatrix} 1 & b \\ 0 & 1 \end{pmatrix} x \begin{pmatrix} \cos\phi \\ -\frac{k}{2}\sin\phi \end{pmatrix} = x \begin{pmatrix} \cos\phi - \frac{kb}{2}\sin\phi \\ -\frac{k}{2}\sin\phi \end{pmatrix} = \begin{pmatrix} 0 \\ x'_f \end{pmatrix}.$$ (4.229)

Solving for b yields

$$b = \frac{2}{k} \cot \phi, \tag{4.230}$$

and the effective focal length f must be

$$f = -\frac{x}{x'_f} = \frac{2}{k \sin \phi}. \tag{4.231}$$

Solving for h we have

$$h = f - b - \frac{l}{2}$$

$$= \frac{2}{k \sin \phi} - \frac{2 \cos \phi}{k \sin \phi} - \frac{l}{2}$$

$$= \frac{2}{k} \frac{\sin^2 \frac{\phi}{2}}{\sin \frac{\phi}{2} \cos \frac{\phi}{2}} - \frac{l}{2}$$

$$= \frac{l}{2} \left(\frac{2}{\phi} \tan \frac{\phi}{2} - 1 \right). \tag{4.232}$$

Reversing the problem, by starting from an upstream "focal" point with

$$\begin{pmatrix} x_i \\ x'_i \end{pmatrix} = \begin{pmatrix} 0 \\ \frac{k}{2} \sin \phi \end{pmatrix}, \tag{4.233}$$

the outgoing ray would now be parallel to the axis with the effective kick located at distance h past the center of the magnet. So the thick lens element may be decomposed with drifts of length

$$h + l/2 = \frac{l}{\phi} \tan \frac{\phi}{2}, \tag{4.234}$$

before and after a thin lens:

$$\mathbf{F} = \begin{pmatrix} 1 & \frac{l}{\phi} \tan \frac{\phi}{2} \\ 0 & 1 \end{pmatrix} \begin{pmatrix} 1 & 0 \\ -\frac{k}{2} \sin \phi & 1 \end{pmatrix} \begin{pmatrix} 1 & \frac{l}{\phi} \tan \frac{\phi}{2} \\ 0 & 1 \end{pmatrix}$$

$$= \begin{pmatrix} 1 - \frac{kl}{2\phi} \tan \frac{\phi}{2} \sin \phi & \frac{l}{\phi} \tan \frac{\phi}{2} \\ -\frac{k}{2} \sin \phi & 1 \end{pmatrix} \begin{pmatrix} 1 & \frac{l}{\phi} \tan \frac{\phi}{2} \\ 0 & 1 \end{pmatrix} \tag{4.235}$$

$$F_{11} = F_{22} = 1 - \frac{kl}{2\phi} \tan \frac{\phi}{2} \sin \phi = 1 - 2 \sin^2 \frac{\phi}{2} = \cos \phi, \tag{4.236}$$

$$F_{12} = \left(2 - \frac{kl}{2\phi} \tan \frac{\phi}{2} \sin \phi \right) \frac{l}{\phi} \tan \frac{\phi}{2}$$

$$= 2 \left(1 - \sin^2 \frac{\phi}{2} \right) \frac{k}{2} \tan \frac{\phi}{2} = \frac{k}{2} \sin \phi, \tag{4.237}$$

$$F_{21} = -\frac{k}{2} \sin \phi, \tag{4.238}$$

which are just the elements in Eq. (4.227)

Chapter 5

Problems of Chapter 5: Strong Focusing

5.1 Problem 5–1: Twiss version of de Moivre's theorem

Using Eq. (CM: 5.21), evaluate $\mathbf{M}^k$, $\mathbf{M}^{-1}$, $\mathbf{M}\mathbf{M}^{-1}$, and $\mathbf{M}_1\mathbf{M}_2$, where $\mathbf{M}_1$ and $\mathbf{M}_2$ have different phase advances but the same Twiss parameters.

Solution:

Recall Eq. (CM: 5.21):

$$\mathbf{M} = \begin{pmatrix} \cos\mu + \alpha\sin\mu & \beta\sin\mu \\ -\gamma\sin\mu & \cos\mu - \alpha\sin\mu \end{pmatrix}$$
$$= \begin{pmatrix} 1 & 0 \\ 0 & 1 \end{pmatrix}\cos\mu + \begin{pmatrix} \alpha & \beta \\ -\gamma & -\alpha \end{pmatrix}\sin\mu, \tag{5.1}$$

This is more succinctly written as

$$\mathbf{M} = \mathbf{I}\cos\mu + \mathbf{J}\sin\mu, \quad \text{with the definition} \quad \mathbf{J} = \begin{pmatrix} \alpha & \beta \\ -\gamma & -\alpha \end{pmatrix}. \tag{5.2}$$

We should first start by proving that $\mathbf{M} = e^{\mathbf{J}\mu}$:

$$e^{\mathbf{J}\mu} = \sum_{n=0}^{\infty} \mathbf{J}^n\frac{\mu^n}{n!} = \mathbf{I}\left(1 - \frac{\mu^2}{2!} + \frac{\mu^4}{4!} - \cdots\right) + \mathbf{J}\left(\mu - \frac{\mu^3}{3!} + \frac{\mu^5}{5!} - \cdots\right)$$
$$= \mathbf{I}\cos\mu + \mathbf{J}\sin\mu = \mathbf{M},$$

$$\tag{5.3}$$

since $\mathbf{J}^2 = -\mathbf{I}$. So we get

$$\mathbf{M}^k = (e^{\mathbf{J}\mu})^k = e^{\mathbf{J}k\mu} = \mathbf{I}\cos(k\mu) + \mathbf{J}\sin(k\mu). \tag{5.4}$$

This is actually just de Moivre's theorem, since $\mathbf{J}$ is a particular 2×2 real matrix representation of $i = \sqrt{-1}$.

The inverse of the matrix is

$$
\begin{aligned}
\mathbf{M}^{-1} &= \begin{pmatrix} \cos\mu - \alpha\sin\mu & -\beta\sin\mu \\ \gamma\sin\mu & \cos\mu + \alpha\sin\mu \end{pmatrix} \\
&= \begin{pmatrix} 1 & 0 \\ 0 & 1 \end{pmatrix} \cos\mu + \begin{pmatrix} -\alpha & -\beta \\ \gamma & \alpha \end{pmatrix} \sin\mu, \\
&= \mathbf{I}\cos\mu - \mathbf{J}\sin\mu = e^{-\mathbf{J}\mu}, \quad\quad\quad (5.5)
\end{aligned}
$$

as we should expect, and of course

$$
\begin{aligned}
\mathbf{M}\mathbf{M}^{-1} &= (\mathbf{I}\cos\mu + \mathbf{J}\sin\mu)(\mathbf{I}\cos\mu - \mathbf{J}\sin\mu) \\
&= \mathbf{I}\cos^2\mu - \mathbf{J}^2\sin^2\mu = \mathbf{I}(\cos^2\mu + \sin^2\mu) = \mathbf{I}. \quad (5.6)
\end{aligned}
$$

For $\mathbf{M}_1 = e^{\mathbf{J}\mu_1}$ and $\mathbf{M}_2 = e^{\mathbf{J}\mu_2}$ we get

$$
\mathbf{M}_1\mathbf{M}_2 = e^{\mathbf{J}\mu_1}e^{\mathbf{J}\mu_2} = e^{\mathbf{J}(\mu_1+\mu_2)}, \quad\quad\quad (5.7)
$$

since $\mathbf{J}$ commutes with itself. (See Eq. (CM: 3.116).)

Comments:

That $\mathbf{M}^k = e^{\mathbf{J}k\mu}$ is rather obvious if one considers $\mathbf{M}$ as a periodic cell with $\mu = \int_{\text{cell}} \frac{ds}{\beta(s)}$ integrated over the length of the periodic cell. Repeating this k times will then give $k\mu$ for the phase advance over k periodic cells or k turns in the case of a 1-turn matrix.

The last case demonstrates how an insertion section can easily be placed between periodic cells provided that the lattice parameters are matched at each end. This will be discussed more in the next chapter.

5.2 Problem 5–2: Hill's equations from Hamiltonian

Show that Eqs. (CM: 5.23 and CM: 5.24)

$$x'' + k_x(s)x = \frac{\delta}{\rho(s)}, \qquad \text{(CM: 5.23)}$$

$$y'' + k_y(s)y = 0, \qquad \text{(CM: 5.24)}$$

may be obtained from the Hamiltonian of Eq. (CM: 3.83)

$$\mathcal{H} = -\frac{q}{p_0}A_s - \left(1 + \frac{x}{\rho}\right)\left(1 + \delta - \frac{1}{2}(x'^2 + y'^2) + \dots\right) + 1 + \delta. \quad \text{(CM: 3.83)}$$

Solution:

Expanding the Hamiltonian $\mathcal{H}$ and keeping terms to second order, we find

$$\mathcal{H}(x, x', y, y', z, \delta; s) \simeq -\frac{q}{p_0}A_s - \frac{x}{\rho} - \frac{x\delta}{\rho} + \frac{1}{2}x'^2 + \frac{1}{2}y'^2 + \mathcal{O}(3). \quad (5.8)$$

Application of Hamilton's equations of motion for transverse coordinates produces

$$\frac{dx}{ds} = \frac{\partial \mathcal{H}}{\partial x'} \simeq x', \qquad (5.9)$$

$$x'' = \frac{dx'}{ds} = -\frac{\partial \mathcal{H}}{\partial x} \simeq \frac{q}{p_0}\frac{\partial A_s}{\partial x} + \frac{1}{\rho} + \frac{\delta}{\rho}, \qquad (5.10)$$

$$\frac{dy}{ds} = \frac{\partial \mathcal{H}}{\partial y'} \simeq y', \qquad (5.11)$$

$$y'' = \frac{dy'}{ds} = -\frac{\partial \mathcal{H}}{\partial y} \simeq \frac{q}{p_0}\frac{\partial A_s}{\partial y}. \qquad (5.12)$$

Recalling Eqs. (CM: 3.48 and CM: 3.49) for the transverse B field components when the vector potential is purely longitudinal,

$$B_x = \frac{1}{1 + x/\rho}\frac{\partial A_s}{\partial y}, \qquad \text{(CM: 3.48)}$$

$$B_y = -\frac{1}{1 + x/\rho}\frac{\partial A_s}{\partial x}, \qquad \text{(CM: 3.49)}$$

and solving for the gradients of A_s gives

$$\frac{\partial A_s}{\partial x} = -\left(1 + \frac{x}{\rho}\right)B_y = -\left(1 + \frac{x}{\rho}\right)(B_0 + gx + \cdots), \qquad (5.13)$$

$$\frac{\partial A_s}{\partial y} = \left(1 + \frac{x}{\rho}\right)B_x = \left(1 + \frac{x}{\rho}\right)(gy + \cdots), \qquad (5.14)$$

where the transverse gradients on the design orbit are

$$g = \left(\frac{\partial B_x}{\partial y}\right)_0 = \left(\frac{\partial B_y}{\partial x}\right)_0.$$ (5.15)

Substituting Eq. (5.13) into Eq. (5.10) yields

$$\begin{aligned}
x'' &= \frac{1}{\rho} - \frac{qB_0}{\rho} - \frac{qB_0}{p_0\rho}\frac{x}{\rho} - \frac{qg}{p_0}x + \frac{\delta}{\rho} \\
&= -\left[\frac{1}{\rho^2} + \frac{q}{p_0}\left(\frac{\partial B_y}{\partial x}\right)_0\right]x + \frac{\delta}{\rho},
\end{aligned}$$ (5.16)

since the cyclotron radius is

$$\rho = \frac{p_0}{qB_0}.$$ (5.17)

Note that the terms inside the brackets are

$$k_x(s) = \frac{1}{\rho^2} + \frac{q}{p_0}\left(\frac{\partial B_y}{\partial x}\right)_0,$$ (5.18)

which is identical to Eq. (CM: 5.25), so after moving the x term to the left hand side of Eq. (5.16), we have

$$x'' + k_x(s)\,x = \frac{\delta}{\rho(s)}.$$ (CM: 5.23)

In a similar manner Eqs. (5.12 and 5.14) combine to make

$$y'' = \frac{qg}{p_0}y \quad \text{or} \quad y'' + k_y(s)y = 0,$$ (5.19)

with

$$k_y(s) = -\frac{q}{p_0}\left(\frac{\partial B_y}{\partial x}\right)_0.$$ (5.20)

5.3 Problem 5–3: Courant-Snyder ellipse properties

a) Show that the equation

$$\gamma x^2 + 2\alpha xy + \beta y^2 = \mathcal{W},\tag{5.21}$$

with $\beta\gamma - \alpha^2 = 1$ is the equation of an ellipse, centered on the origin.

b) Show that the area of this ellipse is given by $A = \pi\mathcal{W}$.

c) Show that

$$\begin{array}{lll}\text{i)} & x_{\max} = \sqrt{\beta\mathcal{W}} & \text{when } y = -\alpha\sqrt{\mathcal{W}/\beta},\\\text{ii)} & y_{\max} = \sqrt{\gamma\mathcal{W}} & \text{when } x = -\alpha\sqrt{\mathcal{W}/\gamma},\\\text{iii)} & y_0 = \sqrt{\mathcal{W}/\beta} & \text{when } x = 0,\\\text{and} \quad \text{iv)} & x_0 = \sqrt{\mathcal{W}/\gamma} & \text{when } y = 0.\end{array}$$

Note that the area $A = \pi x_0 y_{\max} = \pi y_0 x_{\max}$.

Solution:

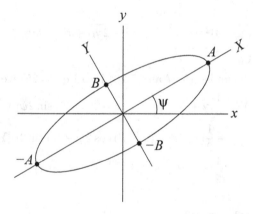

Fig. 5.1 Rotation by an angle ψ from a frame where the ellipse is skew to a frame where the major and minor axes of the ellipse are aligned to the frame's axes.

a) Let us attempt to transform Eq. (5.21) into the form

$$\left(\frac{X}{a}\right)^2 + \left(\frac{X}{b}\right)^2 = 1,\tag{5.22}$$

by rotating our coordinate system as shown in Fig. 5.1:

$$\begin{pmatrix} x \\ y \end{pmatrix} = \begin{pmatrix} \cos\psi & -\sin\psi \\ \sin\psi & \cos\psi \end{pmatrix} \begin{pmatrix} X \\ Y \end{pmatrix}.\tag{5.23}$$

Inserting this into Eq. (5.21) yields

$$\mathcal{W} = (\gamma \cos^2 \psi + \beta \sin^2 \psi + 2\alpha \sin \psi \cos \psi) X^2$$
$$+ (\gamma \sin^2 \psi + \beta \cos^2 \psi - 2\alpha \sin \psi \cos \psi) Y^2$$
$$+ 2[\alpha(\cos^2 \psi - \sin^2 \psi) - (\gamma - \beta) \sin \psi \cos \psi] XY, \quad (5.24)$$

which reduces to

$$\mathcal{W} = (\gamma \cos^2 \psi + \beta \sin^2 \psi + \alpha \sin 2\psi) X^2 + (\gamma \sin^2 \psi + \beta \cos^2 \psi - \alpha \sin 2\psi) Y^2, \quad (5.25)$$

if the coefficient $[2\alpha \cos 2\psi - (\gamma - \beta) \sin 2\psi]$ of XY is null, i. e. when

$$\tan 2\psi = \frac{2\alpha}{\gamma - \beta}, \quad (5.26)$$

or equivalently

$$\sin 2\psi = \tan 2\psi [1 + \tan^2 2\psi]^{-\frac{1}{2}} = \frac{2\alpha}{S}, \quad (5.27)$$

$$\cos 2\psi = [1 + \tan^2 2\psi]^{-\frac{1}{2}} = \frac{\gamma - \beta}{S}, \quad (5.28)$$

with

$$S = \sqrt{(\gamma - \beta)^2 + 4\alpha^2} = \sqrt{\gamma^2 - 2\gamma\beta + \beta^2 + 4(\gamma\beta - 1)}$$
$$= \sqrt{(\gamma + \beta)^2 - 4}. \quad (5.29)$$

Inserting the formulae (5.27 and 5.28) into Eq. (5.25) we obtain:

$$\mathcal{W} = \frac{1}{2}[\gamma + \beta + (\gamma - \beta) \cos 2\psi + 2\alpha \sin 2\psi] X^2$$
$$+ \frac{1}{2}[\gamma + \beta - (\gamma - \beta) \cos 2\psi - 2\alpha \sin 2\psi] Y^2$$
$$= \frac{1}{2}(\gamma + \beta + S) X^2 + \frac{1}{2}(\gamma + \beta - S) Y^2, \quad (5.30)$$

or

$$\mathcal{W} = C_+ X^2 + C_- Y^2 \implies \frac{X^2}{\left(\sqrt{\frac{\mathcal{W}}{C_+}}\right)^2} + \frac{Y^2}{\left(\sqrt{\frac{\mathcal{W}}{C_-}}\right)^2} = 1, \quad (5.31)$$

with the definitions

$$C_\pm = \frac{1}{2}(\beta + \gamma \pm S). \quad (5.32)$$

Eq. (5.31) will be an ellipse of the form given in Eq. (5.22), provided that $C_\pm > 0$. Since β is always positive, then $\gamma = (1 + \alpha^2)/\beta > 0$, and furthermore,

$$\beta + \gamma = \beta + \frac{1 + \alpha^2}{\beta} \geq \beta + \frac{1}{\beta} \geq 2; \quad (5.33)$$

therefore

$$\beta + \gamma > S \geq 0, \tag{5.34}$$

which implies that $C_\pm > 0$, as required. So Eq. (5.21) really is the equation of an ellipse with major an minor radii

$$a = \sqrt{W/C_+}, \quad \text{and} \quad b = \sqrt{W/C_-}. \tag{5.35}$$

By symmetry, it is obvious that the ellipse's center is at the origin.

b) The area of this ellipse is

$$A = \pi a b = \frac{\pi W}{\sqrt{C_+ C_-}} = \frac{4\pi W}{(\beta+\gamma)^2 - [(\gamma+\beta)^2 - 4]} = \pi W. \tag{5.36}$$

c) i) First, for the case $\alpha = 0$, we have $\gamma = 1/\beta$, and

$$x^2 = \beta(W - y^2), \tag{5.37}$$

so it is obvious that

$$x_{max} = \sqrt{\beta W}, \quad \text{when} \quad y = 0. \tag{5.38}$$

When $\alpha \neq 0$, take the differential of Eq. (5.21):

$$2(\gamma x + \alpha y)dx + 2(\alpha x + \beta y)dy = 0. \tag{5.39}$$

For an extremum we require that

$$\frac{dx}{dy} = -\frac{\alpha x + \beta y}{\gamma x + \alpha y} = 0, \tag{5.40}$$

so x_{max} will lie on the line:

$$x = -\frac{\beta}{\alpha}y. \tag{5.41}$$

The intersection of the line and ellipse will have

$$W = \left(\gamma \frac{\beta^2}{\alpha^2} - 2\beta + \beta\right) y^2 = (\gamma\beta - \alpha^2)\beta y^2 = \frac{\beta}{\alpha^2}y^2, \quad \text{or} \tag{5.42}$$

$$y = \pm\alpha\sqrt{\frac{W}{\beta}}, \tag{5.43}$$

so x will have a maximum at

$$(x_{max}, y) = (\sqrt{W\beta}, -\alpha\sqrt{W/\beta}). \tag{5.44}$$

ii) By the symmetry of Eq. (5.21), if we swap x and y simultaneously while swapping β and γ, we find the maximum of y at

$$(x, y_{max}) = (-\alpha\sqrt{W/\gamma}, \sqrt{W\gamma}). \tag{5.45}$$

iii) Setting $x = 0$ in Eq. (5.21), we have $y_0 = \pm\sqrt{W\beta}$.

iv) Again using the symmetry, we have for $y = 0$: $x_0 = \pm\sqrt{W\gamma}$.

5.4   Problem 5-4: Floquet transformation to a circle

a) Show that the transformation

$$
\begin{pmatrix} \xi \\ \zeta \end{pmatrix} = \begin{pmatrix} \beta^{-\frac{1}{2}} & 0 \\ \alpha\beta^{-\frac{1}{2}} & \beta^{\frac{1}{2}} \end{pmatrix} \begin{pmatrix} z \\ z' \end{pmatrix}, \tag{5.46}
$$

transforms the transfer matrix,

$$
\mathbf{M} = e^{\mathbf{J}\mu}, \tag{5.47}
$$

into the matrix,

$$
\mathbf{N} = \begin{pmatrix} \cos\mu & \sin\mu \\ -\sin\mu & \cos\mu \end{pmatrix}. \tag{5.48}
$$

These new coordinates (ξ, ζ) are sometimes referred to as Floquet or Courant-Snyder coordinates. Note that the ellipse of the Courant-Snyder invariant has been transformed to a circle. Show that the invariant remains unchanged by this transformation.

b) Consider a Gaussian distribution of particles in the new coordinates,

$$
f = \frac{N}{2\pi\epsilon} \exp\left(-\frac{\xi^2 + \zeta^2}{2\epsilon} \right). \tag{5.49}
$$

Find the distribution in the old coordinates (z, z'). Evaluate the variances $\sigma_z^2 = \langle (z - \langle z \rangle)^2 \rangle$, and $\sigma_{z'}^2 = \langle (z' - \langle z' \rangle)^2 \rangle$, and the covariance $\sigma_{zz'}^2 = \langle (z - \langle z \rangle)(z' - \langle z' \rangle) \rangle$.

Solution:

a) Let us consider the transformation

$$
\begin{pmatrix} \xi \\ \zeta \end{pmatrix} = \begin{pmatrix} \beta^{-\frac{1}{2}} & 0 \\ \alpha\beta^{-\frac{1}{2}} & \beta^{\frac{1}{2}} \end{pmatrix} \begin{pmatrix} z \\ z' \end{pmatrix} = \mathbf{Q} \begin{pmatrix} z \\ z' \end{pmatrix} = \begin{pmatrix} \beta^{-\frac{1}{2}} z \\ \alpha\beta^{-\frac{1}{2}} z + \beta^{\frac{1}{2}} z' \end{pmatrix} \tag{5.50}
$$

and its inverse

$$
\begin{pmatrix} z \\ z' \end{pmatrix} = \begin{pmatrix} \beta^{\frac{1}{2}} & 0 \\ -\alpha\beta^{-\frac{1}{2}} & \beta^{-\frac{1}{2}} \end{pmatrix} \begin{pmatrix} \xi \\ \zeta \end{pmatrix} = \mathbf{Q}^{-1} \begin{pmatrix} \xi \\ \zeta \end{pmatrix} = \begin{pmatrix} \beta^{\frac{1}{2}} \zeta \\ -\alpha\beta^{-\frac{1}{2}} \xi + \beta^{-\frac{1}{2}} \zeta \end{pmatrix}, \tag{5.51}
$$

having considered the properties of any 2×2 matrix and its inverse

$$
\mathbf{H} = \begin{pmatrix} H_{11} & H_{12} \\ H_{21} & H_{22} \end{pmatrix}, \quad \Longrightarrow \quad \mathbf{H}^{-1} = \frac{1}{\det(\mathbf{H})} \begin{pmatrix} H_{22} & -H_{12} \\ -H_{21} & H_{11} \end{pmatrix}, \tag{5.52}
$$

together with the fact that the determinant of both the 2×2 matrices is $\beta^{-\frac{1}{2}} \beta^{\frac{1}{2}} = 1$. Since the determinant of $\mathbf{Q}$ is 1, the area of an ellipse in the (z, z') coordinates as specified by the Courant-Snyder invariant will remain unchanged when the coordinates are transformed to the (ξ, ζ) plane.

The matrix to be transformed is

$$\mathbf{M} = \begin{pmatrix} \cos\mu + \alpha\sin\mu & \beta\sin\mu \\ -\gamma\sin\mu & \cos\mu - \alpha\sin\mu \end{pmatrix}$$

$$= \begin{pmatrix} 1 & 0 \\ 0 & 1 \end{pmatrix} \cos\mu + \begin{pmatrix} \alpha & \beta \\ -\gamma & -\alpha \end{pmatrix} \sin\mu \qquad \text{(CM:5.21)}$$

$$= \mathbf{I}\cos\mu + \mathbf{J}\sin\mu = e^{\mathbf{J}\mu}. \qquad \text{(CM: 5.22)}$$

This matrix can be used for turning a vector $\begin{pmatrix} z_0 \\ z_0' \end{pmatrix}$ into a vector $\begin{pmatrix} z \\ z' \end{pmatrix}$; namely

$$\begin{pmatrix} z \\ z' \end{pmatrix} = \mathbf{M} \begin{pmatrix} z_0 \\ z_0' \end{pmatrix}. \qquad (5.53)$$

Transforming to the (ξ, ζ) system via Eq. (5.51):

$$\mathbf{Q}^{-1} \begin{pmatrix} \xi \\ \zeta \end{pmatrix} = \mathbf{M}\mathbf{Q}^{-1} \begin{pmatrix} \xi_0 \\ \zeta_0 \end{pmatrix}, \qquad (5.54)$$

or after rearrangement:

$$\begin{pmatrix} \xi \\ \zeta \end{pmatrix} = \mathbf{Q}\mathbf{M}\mathbf{Q}^{-1} \begin{pmatrix} \xi_0 \\ \zeta_0 \end{pmatrix} = \mathbf{N} \begin{pmatrix} \xi_0 \\ \zeta_0 \end{pmatrix}, \qquad (5.55)$$

with

$$\mathbf{N} = \mathbf{I}\cos\mu + \mathbf{Q}\mathbf{J}\mathbf{Q}^{-1}\sin\mu. \qquad (5.56)$$

$$\mathbf{Q}\mathbf{J}\mathbf{Q}^{-1} = \begin{pmatrix} \beta^{-\frac{1}{2}} & 0 \\ \alpha\beta^{-\frac{1}{2}} & \beta^{\frac{1}{2}} \end{pmatrix} \begin{pmatrix} \alpha & \beta \\ -\gamma & -\alpha \end{pmatrix} \begin{pmatrix} \beta^{\frac{1}{2}} & 0 \\ -\alpha\beta^{-\frac{1}{2}} & \beta^{-\frac{1}{2}} \end{pmatrix}$$

$$= \begin{pmatrix} \alpha\beta^{-\frac{1}{2}} & \beta^{\frac{1}{2}} \\ \alpha^2\beta^{-\frac{1}{2}} - \beta^{\frac{1}{2}}\gamma & 0 \end{pmatrix} \begin{pmatrix} \beta^{\frac{1}{2}} & 0 \\ -\alpha\beta^{-\frac{1}{2}} & \beta^{-\frac{1}{2}} \end{pmatrix} = \begin{pmatrix} 0 & 1 \\ -1 & 0 \end{pmatrix}. \quad (5.57)$$

Hence we obtain

$$\mathbf{N} = \begin{pmatrix} 1 & 0 \\ 0 & 1 \end{pmatrix} \cos\mu + \begin{pmatrix} 0 & 1 \\ -1 & 0 \end{pmatrix} \sin\mu = \begin{pmatrix} \cos\mu & \sin\mu \\ -\sin\mu & \cos\mu \end{pmatrix}. \qquad (5.58)$$

Let us recall the Courant-Snyder invariant

$$\gamma z^2 + 2\alpha z z' = \beta z'^2 = \mathcal{W}, \tag{5.59}$$

and let us implement the transformation $(z, z') \to (\xi, \zeta)$ described by Eq. (5.50):

$$
\begin{aligned}
\mathcal{W} &= \gamma(\beta \xi^2) + 2\alpha \left(\beta^{\frac{1}{2}} \xi\right) \left(-\alpha \beta^{-\frac{1}{2}} \xi + \beta^{-\frac{1}{2}} \zeta\right) + \beta \left(\frac{(\zeta - \alpha \xi)^2}{\beta}\right) \\
&= \beta \gamma \xi^2 - 2\alpha^2 \xi^2 + 2\alpha \zeta \xi + \zeta^2 - 2\alpha \xi \zeta + \alpha^2 \xi^2 \\
&= (\beta \gamma - \alpha^2) \xi^2 + \zeta^2 \\
&= \xi^2 + \zeta^2,
\end{aligned}
\tag{5.60}
$$

which is the equation of a circle of radius $r = \sqrt{\mathcal{W}}$.

b) We have for our 2-d Gaussian distribution of particles:

$$f(\xi, \zeta)\, d\xi\, d\zeta = \frac{N}{2\pi\epsilon} \exp\left(-\frac{\xi^2 + \zeta^2}{2\epsilon}\right) d\xi\, d\zeta. \tag{5.61}$$

Transforming back to the original (z, z') coordinates:

$$f(z, z')\, dz\, dz' = \frac{N}{2\pi\epsilon} \exp\left(-\frac{\gamma z^2 + 2\alpha z z' + \beta z'^2}{2\epsilon}\right) dz\, dz'. \tag{5.62}$$

Integrating Eq. (5.61) we have

$$
\begin{aligned}
\frac{N}{2\pi\epsilon} \int_{-\infty}^{\infty} \int_{-\infty}^{\infty} \exp\left(-\frac{\xi^2 + \zeta^2}{2\epsilon}\right) d\xi\, d\zeta &= \frac{N}{2\pi\epsilon} \int_{0}^{2\pi} \int_{0}^{\infty} \exp\left(-\frac{r^2}{2\epsilon}\right) r\, dr\, d\theta \\
&= \frac{N}{\epsilon} \int_{0}^{\infty} \exp\left(-\frac{r^2}{2\epsilon}\right) d\left(\frac{r^2}{2\epsilon}\right) \epsilon = -N \exp\left(-\frac{r^2}{2\epsilon}\right)\Big|_{0}^{\infty} = N,
\end{aligned}
\tag{5.63}
$$

where N is the number of particles in the beam.

It will be convenient to rearrange the Courant-Snyder invariant expression in Eq. (5.62) by completing the square of the quadratic form:

$$
\begin{aligned}
\gamma z^2 + 2\alpha z z' + \beta z'^2 &= \gamma \left(z + \frac{\alpha z'}{\gamma}\right)^2 + \beta z'^2 - \frac{\alpha^2}{\gamma} z'^2 \\
&= \gamma \left(z + \frac{\alpha z'}{\gamma}\right)^2 + \frac{z'^2}{\gamma}.
\end{aligned}
\tag{5.64}
$$

We wish to calculate the variances: $\langle z \rangle$, $\langle z' \rangle$, $\langle z^2 \rangle$, $\langle z'^2 \rangle$, and $\langle z z' \rangle$.

For the average position:

$$\langle z \rangle = \frac{1}{N} \int_{-\infty}^{\infty} \int_{-\infty}^{\infty} z f \, dz \, dz'$$

$$= \frac{1}{2\pi\epsilon} \int_{-\infty}^{\infty} \int_{-\infty}^{\infty} z \exp\left(-\frac{\gamma(z + \alpha z'/\gamma)^2 + z'^2/\gamma}{2\epsilon}\right) dz \, dz'$$

$$= \frac{1}{2\pi\epsilon} \int_{-\infty}^{\infty} \exp\left(-\frac{z'^2}{2\epsilon\gamma}\right) \left[\int_{-\infty}^{\infty} z \exp\left(-\frac{\gamma(z + \alpha z'/\gamma)^2}{2\epsilon}\right) dz\right] dz'$$

$$= \frac{1}{2\pi\epsilon} \int_{-\infty}^{\infty} \exp\left(-\frac{z'^2}{2\epsilon\gamma}\right) \left[\int_{-\infty}^{\infty} z \exp\left(-\frac{\gamma z^2}{2\epsilon}\right) dz\right] dz'$$

$$= 0, \tag{5.65}$$

since the integrand in the inner integral is an odd function of z. This should be expected by symmetry, since we started with a centered Gaussian distribution. Due to the symmetry of the Courant-Snyder invariant expression inside the exponential of Eq. (5.62), we may evaluate $\langle z' \rangle$ by interchanging both β and γ simultaneously with the pair z and z' in the expression for $\langle z \rangle$ to find also that

$$\langle z' \rangle = 0. \tag{5.66}$$

The variance $\sigma_z^2 = \langle (z - \langle z \rangle)^2 \rangle = \langle z^2 \rangle$ since $\langle z \rangle = 0$ is

$$\langle z^2 \rangle = \frac{1}{2\pi\epsilon} \int_{-\infty}^{\infty} \exp\left(-\frac{z'^2}{2\epsilon\gamma}\right) \left[\int_{-\infty}^{\infty} z^2 \exp\left(-\frac{\gamma(z + \alpha z'/\gamma)^2}{2\epsilon}\right) dz\right] dz'$$

$$= \frac{1}{2\pi\epsilon} \int_{-\infty}^{\infty} \exp\left(-\frac{z'^2}{2\epsilon\gamma}\right) \left[\int_{-\infty}^{\infty} \left(y - \frac{\alpha z'}{\gamma}\right)^2 \exp\left(-\frac{\gamma y^2}{2\epsilon}\right) dy\right] dz', \tag{5.67}$$

with the substitution $y = z + \alpha z'/\gamma$. Let us first calculate the inner integral:

$$I_1 = \int_{-\infty}^{\infty} \left(y - \frac{\alpha z'}{\gamma}\right)^2 \exp\left(-\frac{\gamma y^2}{2\epsilon}\right) dy$$

$$= \int_{-\infty}^{\infty} \left(y^2 - 2\frac{\alpha y z'}{\gamma} + \frac{\alpha^2 z'^2}{\gamma^2}\right) \exp\left(-\frac{\gamma y^2}{2\epsilon}\right) dy$$

$$= \int_{-\infty}^{\infty} y^2 \exp\left(-\frac{\gamma y^2}{2\epsilon}\right) dy + \frac{\alpha^2 z'^2}{\gamma^2} \int_{-\infty}^{\infty} \exp\left(-\frac{\gamma y^2}{2\epsilon}\right) dy. \tag{5.68}$$

Recall that for a Gaussian distribution, we have the integrals:

$$\frac{1}{\sqrt{2\pi}\,\sigma} \int_{-\infty}^{\infty} \exp\left(-\frac{x^2}{2\sigma^2}\right) dx = 1,$$

$$\frac{1}{\sqrt{2\pi}\,\sigma} \int_{-\infty}^{\infty} x^2 \exp\left(-\frac{x^2}{2\sigma^2}\right) dx = \sigma^2. \tag{5.69}$$

So with $\sigma = \sqrt{\frac{\epsilon}{\gamma}}$, I_1 becomes

$$I_1 = \sqrt{2\pi}\,\sigma^3 + \frac{\alpha^2 z'^2}{\gamma^2}\sqrt{2\pi}\,\sigma$$

$$= \sqrt{\frac{2\pi\epsilon}{\gamma}}\left(\frac{\epsilon}{\gamma} + \frac{\alpha^2}{\gamma^2}z'^2\right). \tag{5.70}$$

Substituting this result back into the outer integral of Eq. (5.67):

$$\langle z^2 \rangle = \frac{1}{2\pi\epsilon}\sqrt{\frac{2\pi\epsilon}{\gamma}}\int_{-\infty}^{\infty}\left(\frac{\epsilon}{\gamma} + \frac{\alpha^2}{\gamma^2}z'^2\right)\exp\left(-\frac{z'^2}{2\gamma\epsilon}\right)dz'$$

$$= \frac{1}{\sqrt{2\pi\gamma\epsilon}}\int_{-\infty}^{\infty}\left(\frac{\epsilon}{\gamma} + \frac{\alpha^2}{\gamma^2}z'^2\right)\exp\left(-\frac{z'^2}{2\gamma\epsilon}\right)dz'$$

$$= \frac{1}{\sqrt{2\pi\gamma\epsilon}}\left(\sqrt{2\pi\gamma\epsilon}\,\frac{\epsilon}{\gamma} + \frac{\alpha^2}{\gamma^2}\sqrt{2\pi}\,(\gamma\epsilon)^{3/2}\right)$$

$$= \epsilon\,\frac{1+\alpha^2}{\gamma}$$

$$= \beta\epsilon. \tag{5.71}$$

Again, by symmetry of $\gamma z^2 + 2\alpha zz' + \beta z'^2$, we can interchange β and γ along with z and z' to get

$$\sigma_{z'}^2 = \langle(z' - \langle z'\rangle)^2\rangle = \langle z'^2\rangle = \gamma\epsilon. \tag{5.72}$$

For the cross-term average:

$$\langle zz'\rangle = \frac{1}{2\pi\epsilon}\int_{-\infty}^{\infty}z'\exp\left(-\frac{z'^2}{2\gamma\epsilon}\right)\left[\int_{-\infty}^{\infty}z\exp\left(-\frac{\gamma(z+\alpha z'/\gamma)^2}{2\epsilon}\right)dz\right]dz'$$

$$= \frac{1}{2\pi\epsilon}\int_{-\infty}^{\infty}z'\exp\left(-\frac{z'^2}{2\gamma\epsilon}\right)\left[\int_{-\infty}^{\infty}\left(y - \frac{\alpha}{\gamma}z'\right)\exp\left(-\frac{\gamma y^2}{2\epsilon}\right)dy\right]dz'$$

$$= -\frac{\alpha}{\gamma}\sqrt{\frac{2\pi\epsilon}{\gamma}}\,\frac{1}{2\pi\epsilon}\int_{-\infty}^{\infty}z'^2\exp\left(-\frac{z'^2}{2\gamma\epsilon}\right)dz'$$

$$= -\frac{\alpha}{\gamma}\,\frac{1}{\sqrt{2\pi\gamma\epsilon}}\sqrt{2\pi}\,(\gamma\epsilon)^{3/2}$$

$$= -\alpha\epsilon. \tag{5.73}$$

Comments:

For the 1-d betatron motion ignoring dispersion, the Σ-matrix of variances of the beam may be written as

$$\Sigma = \langle \mathbf{Z}\mathbf{Z}^{\mathrm{T}} \rangle = \left\langle \begin{pmatrix} z \\ z' \end{pmatrix} \begin{pmatrix} z & z' \end{pmatrix} \right\rangle = \begin{pmatrix} \langle z^2 \rangle & \langle zz' \rangle \\ \langle z'z \rangle & \langle z'^2 \rangle \end{pmatrix} = \begin{pmatrix} \beta\epsilon & -\alpha\epsilon \\ -\alpha\epsilon & \gamma\epsilon \end{pmatrix}, \quad (5.74)$$

where we have written the 2-component vector $\mathbf{Z}$ as

$$\mathbf{Z} = \begin{pmatrix} z \\ z' \end{pmatrix}. \quad (5.75)$$

Note that the determinant of the covariance matrix for a Gaussian distribution is

$$|\Sigma| = \epsilon^2(\beta\gamma - \alpha^2) = \epsilon^2. \quad (5.76)$$

This provides a useful operational way to calculate the emittance of a bunch of particles in tracking simulations. We can calculate the covariance matrix by averaging over all the particles in the simulation, so an estimate of emittance may then be obtained from the square root of the the determinant of the Σ-matrix.

For higher dimensions (2-d or 3-d), we must be more careful in extracting the emittances. In the case where there is dispersion, it might be better to use the inverse of the covariance matrix (frequently referred to as the Fisher matrix in statistics). For the 3-d case with no transverse coupling, the Fisher matrix $\Xi = \Sigma^{-1}$ from (CM: D.22) is

$$\Xi = \begin{pmatrix} \dfrac{\gamma_x}{\epsilon_x} & \dfrac{\alpha_x}{\epsilon_x} & 0 & 0 & 0 & -\dfrac{\gamma_x\eta_x + \alpha_x\eta_x'}{\epsilon_x} \\[2mm] \dfrac{\alpha_x}{\epsilon_x} & \dfrac{\beta_x}{\epsilon_x} & 0 & 0 & 0 & -\dfrac{\alpha_x\eta_x + \beta_x\eta_x'}{\epsilon_x} \\[2mm] 0 & 0 & \dfrac{\gamma_y}{\epsilon_y} & \dfrac{\alpha_y}{\epsilon_y} & 0 & 0 \\[2mm] 0 & 0 & \dfrac{\alpha_y}{\epsilon_y} & \dfrac{\beta_y}{\epsilon_y} & 0 & 0 \\[2mm] 0 & 0 & 0 & 0 & \dfrac{\gamma_z}{\epsilon_z} & \dfrac{\alpha_z}{\epsilon_z} \\[2mm] -\dfrac{\gamma_x\eta_x + \alpha_x\eta_x'}{\epsilon_x} & -\dfrac{\alpha_x\eta_x + \beta_x\eta_x'}{\epsilon_x} & 0 & 0 & \dfrac{\alpha_z}{\epsilon_z} & \dfrac{\beta_z}{\epsilon_z} + \dfrac{H}{\epsilon_x} \end{pmatrix}. \quad \text{(CM:D.22)}$$

Here the upper 2×2-block has the determinant ϵ_x^{-2}, whereas the upper 2×2-block of the Σ-matrix from Eq. (CM: D.24) contains dispersion information:

$$\Sigma_x = \begin{pmatrix} \beta_x\epsilon_x + \eta_x^2\sigma_\delta^2 & -\alpha_x\epsilon_x + \eta_x\eta_x'\sigma_\delta^2 \\ -\alpha_x\epsilon_x + \eta_x\eta_x'\sigma_\delta^2 & \gamma_x\epsilon_x + \eta_x'^2\sigma_\delta^2 \end{pmatrix}. \quad (5.77)$$

5.5   Problem 5–5: Propagation of envelope parameters

Using the invariant
$$\mathcal{W} = \gamma z^2 + 2\alpha z z' + \beta z'^2, \tag{5.78}$$
show that the Twiss parameters transform from s_1 to s_2 by the matrix transformation
$$\begin{pmatrix} \beta_2 \\ \alpha_2 \\ \gamma_2 \end{pmatrix} = \begin{pmatrix} M_{11}^2 & -2M_{11}M_{12} & M_{12}^2 \\ -M_{11}M_{21} & M_{11}M_{22} + M_{12}M_{21} & -M_{12}M_{22} \\ M_{21}^2 & -2M_{21}M_{22} & M_{22}^2 \end{pmatrix} \begin{pmatrix} \beta_1 \\ \alpha_1 \\ \gamma_1 \end{pmatrix}, \tag{5.79}$$

$$\text{if} \quad \begin{pmatrix} z_2 \\ z_2' \end{pmatrix} = \begin{pmatrix} M_{11} & M_{12} \\ M_{21} & M_{22} \end{pmatrix} \begin{pmatrix} z_1 \\ z_1' \end{pmatrix}. \tag{5.80}$$

Solution:

Inverting Eq. (5.80) gives
$$\begin{pmatrix} z_1 \\ z_1' \end{pmatrix} = \begin{pmatrix} M_{22} & -M_{12} \\ -M_{21} & M_{11} \end{pmatrix} \begin{pmatrix} z_2 \\ z_2' \end{pmatrix}, \tag{5.81}$$
The invariant calculated at position 1 is
$$\begin{aligned}
\mathcal{W} &= \gamma_1 z_1^2 + 2\alpha_1 z_1 z_1' + \beta_1 (z_1')^2 \\
&= \gamma_1 (M_{22}z_1 - M_{12}z_1')^2 + 2\alpha_1 (M_{22}z_1 - M_{12}z_1')(-M_{21}z_1 + M_{11}z_1') \\
&\quad + \beta_1 (-M_{21}z_1 + M_{11}z_1')^2 \\
&= z_2^2 [\gamma_1 M_{22}^2 - 2\alpha_1 M_{21}M_{22} + \beta_1 M_{21}^2] \\
&\quad + 2z_2 z_2' [-\gamma_1 M_{12}M_{22} + \alpha_1 (M_{11}M_{22} + M_{12}M_{21}) - \beta_1 M_{11}M_{21}] \\
&\quad + (z_2')^2 [\gamma_1 M_{12}^2 - 2\alpha_1 M_{11}M_{12} + \beta_1 M_{11}^2]. \tag{5.82}
\end{aligned}$$
But since $\mathcal{W}$ is invariant, we also have
$$\mathcal{W} = \gamma_2 z_2^2 + 2\alpha_2 z_2 z_2' + \beta_2 (z_2')^2. \tag{5.83}$$
Comparing the coefficients in the last pair of equations, we have
$$\gamma_2 = \gamma_1 M_{22}^2 - 2\alpha_1 M_{21}M_{22} + \beta_1 M_{21}^2, \tag{5.84}$$
$$\alpha_2 = -\gamma_1 M_{12}M_{22} + \alpha_1 (M_{11}M_{22} + M_{12}M_{21}) - \beta_1 M_{11}M_{21}, \tag{5.85}$$
$$\beta_2 = \gamma_1 M_{12}^2 - 2\alpha_1 M_{11}M_{12} + \beta_1 M_{11}^2, \tag{5.86}$$
which when written in matrix form is Eq. (5.79).

5.6 Problem 5–6: Eigenvalues of the envelope propagation matrix

a) Show that the transfer matrix for the Twiss parameters, given in the Problem 5-5, has a determinant of 1. What does this say about the eigenvalues?

b) Find the eigenvalues and compare the requirements for a stable beam with the results of CM: § 5.2.

Solution:

a) First let us write the matrix in the form

$$\mathbf{U} = \begin{pmatrix} C^2 & -2CS & S^2 \\ -CC' & CS' + C'S & -SS' \\ C'^2 & -2C'S' & S'^2 \end{pmatrix}, \tag{5.87}$$

for the beam transport equation

$$\begin{pmatrix} z_2 \\ z_2' \end{pmatrix} = \begin{pmatrix} C & S \\ C' & S' \end{pmatrix} \begin{pmatrix} z_1 \\ z_1' \end{pmatrix}, \tag{5.88}$$

so that we do not have to keep writing all those tedious subscripts. Since we must find the eigenvalues, it will save us time to first calculate the characteristic polynomial $P(\lambda)$, since the determinant will be simply the value of $P(0)$:

$$
\begin{aligned}
P(\lambda) = |\mathbf{U} - \mathbf{I}\lambda| &= \begin{vmatrix} C^2 - \lambda & -2CS & S^2 \\ -CC' & CS' + C'S - \lambda & -SS' \\ C'^2 & -2C'S' & S'^2 - \lambda \end{vmatrix} \\
&= (C^2 - \lambda)(CS' + C'S - \lambda)(S'^2 - \lambda) + 2CC'^2 S^2 S' + 2CC'^2 S^2 S' \\
&\quad - 2(C^2 - \lambda)C'SS'^2 - 2(S'^2 - \lambda)C^2 C'S - C'^2(CS' + C'S - \lambda)S^2 \\
&= -\lambda^3 + \lambda^2(C^2 + CS' + C'S + S'^2) \\
&\quad - \lambda(C^3 S' + C^2 C'S + C^2 S'^2 + CS'^3 + C'SS'^2 \\
&\qquad - 2C'SS'^2 - 2C^2 C'S - C'^2 S^2) \\
&\quad + C^2(CS' + C'S)S'^2 - C'^2(CS' + C'S)S^2 \\
&\quad + 4CC'^2 S^2 S' - 4C^2 C'SS'^2. \tag{5.89}
\end{aligned}
$$

$$P(\lambda) = -\lambda^3 + \lambda^2(C^2 + 2CS' - 1 + S'^2)$$
$$- \lambda(C^3S' - C^2C'S + C^2S'^2 + CS'^3 - C'SS'^2 - C'^2S^2)$$
$$+ (CS' + C'S)(C^2S'^2 - C'^2S^2) + 4C(C'S)S'(C'S - CS')$$
$$= -\lambda^3 + \lambda^2[(C + S)^2 - 1]$$
$$- \lambda[C^2(CS' - C'S) + (CS' - C'S)S'^2$$
$$+ (CS' - C'S)(CS' + C'S)]$$
$$+ (CS' - C'S)[(CS' + C'S)^2 - 4C(C'S)S']$$
$$= -\lambda^3 + \lambda^2[(C + S)^2 - 1]$$
$$- \lambda(C^2 + S'^2 + CS' + C'S) + (CS' - C'S)^3$$
$$= -\lambda^3 + \lambda^2[(C + S)^2 - 1] - \lambda[(C + S')^2 - 1] + 1$$
$$= 1 - \lambda^3 + \lambda(\lambda - 1)[(C + S')^2 - 1]$$
$$= (1 - \lambda)\{1 + \lambda + \lambda^2 + \lambda[1 - (C + S')^2]\}$$
$$= (1 - \lambda)\{1 + \lambda[2 - (C + S')^2] + \lambda^2\}. \tag{5.90}$$

So clearly $|U| = P(0) = +1$ as desired. Since U is a real 3×3 matrix, it must have one real eigenvalue and the product of the 3 eigenvalues must be 1. The other two eigenvalues may be both real or both complex, but if they are complex, then they must be complex conjugates of each other.

b) Additionally, the characteristic equation $P(\lambda) = 0$ yields one real eigenvalue of $+1$. We also have the equation for the two remaining eigenvalues:

$$\lambda^2 + [2 - (C + S')^2]\lambda + 1 = 0. \tag{5.91}$$

Solving this quadratic equation gives

$$\lambda_\pm = \frac{(C + S')^2}{2} - 1 \pm \sqrt{\left(1 - \frac{(C + S')^2}{2}\right)^2 - 1}. \tag{5.92}$$

For the case of a stable period ring or periodic cell, we have $C + S' = 2\cos\mu$, so the coefficient of λ in the characteristic equation becomes

$$2 - (C + S')^2 = 2 - (2\cos\mu)^2 = 2(1 - 2\cos^2\mu) = 2(\sin^2\mu - \cos^2\mu)$$
$$= -2\cos(2\mu). \tag{5.93}$$

Substituting this into the previous equation, we find the other two eigen-values:

$$\lambda_{\pm} = \cos(2\mu) \pm \sqrt{\cos^2(2\mu) - 1} = \cos(2\mu) \pm i\sin(2\mu)$$
$$= e^{\pm i2\mu}. \tag{5.94}$$

Comment:

Comparing Eq. (5.94) with the eigenvalues of the transport matrix of Eq. (5.88), namely $\lambda_{\pm} = e^{\pm i\mu}$, we see a factor of two in the phase-angle, i. e. the envelope frequency is twice the betatron oscillation frequency. This is related to the fact that an ellipse centered on the origin is identical to the ellipse rotated by 180°, while a single point (or particle trajectory) on the ellipse moves over to the opposite side of the ellipse.

5.7   Problem 5–7: Differential equations of the envelope functions

a) Show that a differential equation for the Twiss parameters may be written in matrix form by

$$
\begin{pmatrix} \beta' \\ \alpha' \\ \gamma' \end{pmatrix} = \begin{pmatrix} 0 & -2 & 0 \\ k & 0 & -1 \\ 0 & 2k & 0 \end{pmatrix} \begin{pmatrix} \beta \\ \alpha \\ \gamma \end{pmatrix}. \tag{5.95}
$$

b) Solve this equation for constant $k \neq 0$. Compare the result with Twiss matrix of Problem (5-5) evaluated for a quadrupole.
c) Solve this equation for $k = 0$. Notice that β is quadratic in free space.
d) Show that

$$
\beta''' + 4k\beta' + 2k'\beta = 0. \tag{5.96}
$$

Solution:

a) From

$$
\alpha = -\frac{1}{2}\beta' = -ww', \tag{CM: 5.47}
$$

we easily see that the first row $\beta' = -2\alpha$ is true. Differentiating

$$
\beta(s) = w^2(s), \tag{CM:5.46}
$$

twice with a few manipulations:

$$
\beta' = 2ww', \tag{5.97}
$$

$$
w' = \frac{\beta'}{2w} = -\frac{\alpha}{\sqrt{\beta}}, \tag{5.98}
$$

$$
\beta'' = 2(w')^2 + 2ww'', \tag{5.99}
$$

$$
\alpha' = -\frac{1}{2}\beta'' = -\left[(w')^2 + ww''\right] = -\left[\frac{\alpha^2}{\beta} + ww''\right]. \tag{5.100}
$$

Multiplying Eq. (CM 5.35): $w'' + kw - w^{-3} = 0$ by w, we find

$$
ww'' = -kw^2 - w^{-2} = -k\beta + \beta^{-1}, \tag{5.101}
$$

and obtain

$$
\alpha' = -\left[\frac{\alpha^2}{\beta} - k\beta + \frac{1}{\beta}\right] = k\beta - \frac{1+\alpha^2}{\beta} = k\beta - \gamma, \tag{5.102}
$$

which matches the second row. Differentiating $\beta\gamma = 1 + \alpha^2$ leads to

$$
\beta\gamma' = -\gamma\beta' + 2\alpha\alpha' = 2\alpha(\gamma + \alpha') = 2\alpha\gamma + 2\alpha(k\beta - \gamma) = 2k\alpha\beta, \tag{5.103}
$$

thus yielding the final row when we divide by β.

b) We can solve the first order differential matrix equation using a Lie algebraic integration as in Chapter 3. Define

$$
\mathbf{G} = 2\sqrt{k}\,\mathbf{K} = 2\sqrt{k}
\begin{pmatrix}
0 & -\frac{1}{\sqrt{k}} & 0 \\
\frac{\sqrt{k}}{2} & 0 & -\frac{1}{2\sqrt{k}} \\
0 & \sqrt{k} & 0
\end{pmatrix}, \quad \text{and} \quad
\mathbf{v} =
\begin{pmatrix}
\beta \\
\alpha \\
\gamma
\end{pmatrix},
$$

(5.104)

so that the differential matrix equation may be written as

$$
\mathbf{v}' = \mathbf{G}\mathbf{v} = 2\sqrt{k}\,\mathbf{K}\mathbf{v}.
$$

(5.105)

Integrating we get

$$
\mathbf{v}(s) = \lim_{n\to\infty} \left(\mathbf{I} + \mathbf{G}\frac{2\sqrt{k}s}{n}\right)^n \mathbf{v}(0) = e^{\mathbf{G}2\sqrt{k}s}\mathbf{v}(0) = e^{\mathbf{G}\phi}\mathbf{v}(0).
$$

(5.106)

with the definition of $\phi = 2\sqrt{k}s$. The characteristic equation for $\mathbf{K}$ is

$$
0 = |\mathbf{K} - \mathbf{I}\lambda| =
\begin{vmatrix}
-\lambda & -\frac{1}{\sqrt{k}} & 0 \\
\frac{\sqrt{k}}{2} & -\lambda & -\frac{1}{2\sqrt{k}} \\
0 & \sqrt{k} & -\lambda
\end{vmatrix}
= -\lambda^3 - \lambda,
$$

(5.107)

so application of the Cayley-Hamilton theorem yields $\mathbf{G}^3 = -\mathbf{G}$. Expanding the exponential yields

$$
e^{\mathbf{G}\phi} = \mathbf{I} + \mathbf{K}\left(\phi - \frac{\phi^3}{3!} + \frac{\phi^5}{5!}\cdots\right) + \mathbf{K}^2\left(\frac{\phi^2}{2!} - \frac{\phi^4}{4!} + \frac{\phi^6}{6!}\cdots\right)
$$

$$
= \mathbf{I} + \mathbf{K}\sin\phi + \mathbf{K}^2(1 - \cos\phi).
$$

(5.108)

The square of $\mathbf{K}$ is

$$
\mathbf{K}^2 =
\begin{pmatrix}
-\frac{1}{2} & 0 & \frac{1}{2k} \\
0 & -1 & 0 \\
\frac{k}{2} & 0 & -\frac{1}{2}
\end{pmatrix},
$$

(5.109)

so the exponential becomes

$$
e^{\mathbf{G}\phi} = \mathbf{I} +
\begin{pmatrix}
0 & -\frac{1}{\sqrt{k}} & 0 \\
\frac{\sqrt{k}}{2} & 0 & -\frac{1}{2\sqrt{k}} \\
0 & \sqrt{k} & 0
\end{pmatrix}\sin\phi +
\begin{pmatrix}
-\frac{1}{2} & 0 & \frac{1}{2k} \\
0 & -1 & 0 \\
\frac{k}{2} & 0 & -\frac{1}{2}
\end{pmatrix}(1 - \cos\phi)
$$

$$
=
\begin{pmatrix}
\frac{1}{2}(1 + \cos\phi) & -\frac{1}{\sqrt{k}}\sin\phi & \frac{1}{2k}(1 - \cos\phi) \\
\frac{\sqrt{k}}{2}\sin\phi & \cos\phi & -\frac{1}{2\sqrt{k}}\sin\phi \\
\frac{k}{2}(1 - \cos\phi) & \sqrt{k}\sin\phi & \frac{1}{2}(1 + \cos\phi)
\end{pmatrix}.
$$

(5.110)

This gives the equation for the transport of Twiss parameters through a quadrupole as

$$\begin{pmatrix} \beta(s) \\ \alpha(s) \\ \gamma(s) \end{pmatrix} = \begin{pmatrix} \frac{1}{2}(1 + \cos\phi) & -\frac{1}{\sqrt{k}}\sin\phi & \frac{1}{2k}(1 - \cos\phi) \\ \frac{\sqrt{k}}{2}\sin\phi & \cos\phi & -\frac{1}{2\sqrt{k}}\sin\phi \\ \frac{k}{2}(1 - \cos\phi) & \sqrt{k}\sin\phi & \frac{1}{2}(1 + \cos\phi) \end{pmatrix} \begin{pmatrix} \beta(0) \\ \alpha(0) \\ \gamma(0) \end{pmatrix}. \quad (5.111)$$

For $k > 0$ we should expect the transport matrix for a focusing quadrupole:

$$\begin{pmatrix} z_2 \\ z_2' \end{pmatrix} = \begin{pmatrix} \cos(\sqrt{k}s) & \frac{1}{\sqrt{k}}\sin(\sqrt{k}s) \\ -\sqrt{k}\sin(\sqrt{k}s) & \cos(\sqrt{k}s) \end{pmatrix} \begin{pmatrix} z_1 \\ z_1' \end{pmatrix}. \quad (5.112)$$

From the previous problem this gives

$$\beta(s) = \beta_1 \cos^2(\sqrt{k}s) - \frac{2\alpha_1}{\sqrt{k}}\sin(\sqrt{k}s)\cos(\sqrt{k}s) + \frac{\gamma_1}{k}\sin^2(\sqrt{k}s)$$

$$= \frac{\beta_1}{2}(1 + \cos\phi) - \frac{\alpha_1}{\sqrt{k}}\sin\phi + \frac{\gamma_1}{2k}(1 - \cos\phi), \quad (5.113)$$

$$\alpha(s) = \beta_1 \sqrt{k}\cos(\sqrt{k}s)\sin(\sqrt{k}s) + \alpha_1 \left[\cos^2(\sqrt{k}s) - \sin^2(\sqrt{k}s)\right]$$

$$\quad - \frac{\gamma_1}{\sqrt{k}}\sin(\sqrt{k}s)\cos(\sqrt{k}s)$$

$$= \frac{\beta_1\sqrt{k}}{2}\sin\phi + \alpha_1\cos\phi - \frac{\gamma_1}{2\sqrt{k}}\sin\phi, \quad (5.114)$$

$$\gamma(s) = \beta_1 k \sin^2(\sqrt{k}s) + 2\alpha_1\sqrt{k}\sin(\sqrt{k}s)\cos(\sqrt{k}s) + \gamma_1\cos^2(\sqrt{k}s)$$

$$= \frac{\beta_1 k}{2}(1 - \cos\phi) + \alpha_1\sqrt{k}\sin\phi + \frac{\gamma_1}{2}(1 + \cos\phi). \quad (5.115)$$

This result agrees with the previous Lie integration.

To get the result for $k < 0$, one can write $\sqrt{k} = i\sqrt{-k}$ and use the usual transformations from trigonometric to hyperbolic functions:

$$\sin\phi = \sin(2i\sqrt{-k}s) = i\sinh(2\sqrt{-k}s), \quad (5.116)$$

$$\cos\phi = \cos(2i\sqrt{-k}s) = \cosh(2\sqrt{-k}s). \quad (5.117)$$

The two methods will still agree, of course. Making these substitutions produces the defocusing quad equation:

$$\begin{pmatrix} \beta(s) \\ \alpha(s) \\ \gamma(s) \end{pmatrix} = \begin{pmatrix} \frac{1}{2}(\cosh\phi + 1) & -\frac{1}{\sqrt{|k|}}\sinh\phi & \frac{1}{2k}(\cosh\phi - 1) \\ -\frac{\sqrt{|k|}}{2}\sinh\phi & \cosh\phi & -\frac{1}{2\sqrt{|k|}}\sinh\phi \\ \frac{k}{2}(\cos\phi - 1) & -\sqrt{|k|}\sinh\phi & \frac{1}{2}(\cosh\phi + 1) \end{pmatrix} \begin{pmatrix} \beta(0) \\ \alpha(0) \\ \gamma(0) \end{pmatrix}, \quad (5.118)$$

with the phase angle ϕ given now by $\phi = 2\sqrt{|k|}s$.

c) For $k = 0$, the equations in reverse order are:

$$\gamma' = 0, \tag{5.119}$$

$$\alpha' = -\gamma, \tag{5.120}$$

$$\beta' = -2\alpha. \tag{5.121}$$

Integrating each in turn leads to

$$\gamma = C_1, \quad \text{a constant}, \tag{5.122}$$

$$\alpha = -C_1 s + C_2, \tag{5.123}$$

$$\beta = -2 \int (-C_1 s + C_2) \, ds$$

$$= C_1 s^2 - 2 C_2 s + C_3, \tag{5.124}$$

where C_2 and C_3 are also constants of integration. This equation is obviously quadratic in s.

d) From part a, we have $\beta' = -2\alpha$, $\alpha' = k\beta - \gamma$, and $\gamma' = 2k\alpha$. Successive differentiation and substitutions lead to

$$\beta'' = -2\alpha' = -2k\beta + 2\gamma, \tag{5.125}$$

$$\beta''' = -2k'\beta - 2k\beta' + 2\gamma'$$

$$= -2k'\beta - 2k\beta' + 2 \left[2k \left(-\frac{1}{2}\beta' \right) \right]$$

$$= -2k'\beta - 4k\beta', \tag{5.126}$$

which is just Eq. (5.96) after terms are brought to one side of the equation.

Comments:

Eq. (5.95) are the Eqs. (3.10, 3.14 and 3.15) of the famous paper[21] of Courant and Snyder, and Eq. (5.96) is also given as Eq. (3.27) of the same reference. Clearly if one starts with Eq. (5.96), then a whole host of initial conditions can be used, but for our purposes we may constrain them to $\beta_0 \gamma_0 - \alpha_0^2 = 1$. This constraint leads to the second order nonlinear differential equation

$$\beta'' = \frac{2 + \alpha^2}{\beta} - 2k\beta = \frac{2 + (\beta'/2)^2}{\beta} - 2k\beta. \tag{5.127}$$

Since, in fact, we derived the differential equation $\gamma' = -2k\alpha$ directly from this constraint in Eq. (5.103), we see that if the initial conditions satisfy the constraint, then the subsequent evolution of β and its derivatives must also satisfy this constraint.

5.8   Comments on the Lie group and algebra of Twiss parameter propagation:

The previous three problems illustrate properties of the transport of Twiss parameters which evoke thoughts of Lie groups and algebras. The 3×3 transport matrices for propagation of the Twiss parameters in fact do form a quadratic Lie group, as mentioned in CM § 3.8.1. In this case, the form of the defining matrix is

$$\mathbf{F} = \begin{pmatrix} 0 & 0 & 1 \\ 0 & -\frac{1}{2} & 0 \\ 1 & 0 & 0 \end{pmatrix}. \tag{5.128}$$

It is easy to verify for a $\mathbf{U}$ of the form in Eq. (5.87), that

$$\mathbf{U}^{-1} = \mathbf{F}\mathbf{U}^{\mathrm{T}}\mathbf{F}^{-1}, \tag{5.129}$$

and hence

$$\mathbf{U}\mathbf{F}\mathbf{U}^{\mathrm{T}} = \mathbf{F} \tag{5.130}$$

by calculating $\mathbf{U}^{-1}$ for the inverse trajectory transport matrix

$$\mathbf{M}^{-1} = \begin{pmatrix} S' & -S \\ -C' & C \end{pmatrix}. \tag{5.131}$$

It is also easy to realize that the all the properties of a group (See Problem 3-3.) are satisfied. Let us call this group G_β. Define a homomorphism from $\mathrm{Sp}(2, \mathbb{R}; \mathbf{S})$ to G_β

$$\Phi(\mathbf{M}) = \begin{pmatrix} M_{11}^2 & -2M_{11}M_{12} & M_{12}^2 \\ -M_{11}M_{21} & M_{11}M_{22} + M_{12}M_{21} & -M_{12}M_{22} \\ M_{21}^2 & -2M_{21}M_{22} & M_{22}^2 \end{pmatrix}. \tag{5.132}$$

For $\mathbf{M}, \mathbf{N} \in \mathrm{Sp}(2, \mathbb{R}; \mathbf{S})$ we will have

$$\Phi(\mathbf{M})\Phi(\mathbf{N}) = \Phi(\mathbf{MN}) \in G_\beta, \tag{5.133}$$

$$\Phi(\mathbf{I}_2) = \mathbf{I}_3, \tag{5.134}$$

$$\Phi(\mathbf{M}^{-1}) = \Phi(\mathbf{M})^{-1}, \tag{5.135}$$

and of course the associative property holds for matrices in general.

If $\mathbf{M} \in \mathrm{Sp}(2, \mathbb{R}; \mathbf{S})$ then so is $-\mathbf{M}$, but clearly $\Phi(-\mathbf{M}) = \Phi(\mathbf{M})$, so G_β is not isomorphic to $\mathrm{Sp}(2, \mathbb{R})$. We can define a quadratic Lie group of 3×3 real matrices by

$$G_{\mathbf{F}} = \{\mathbf{U} \in \mathrm{SL}_3(\mathbb{R}) : \mathbf{U}\mathbf{F}\mathbf{U}^{\mathrm{T}} = \mathbf{F}\}, \tag{5.136}$$

where $SL_3(\mathbb{R})$ is the group of real 3×3 matrices with determinant $+1$. The corresponding Lie algebra will then be defined by

$$\mathfrak{g}_F = \{\mathbf{A} \in \mathbb{C}^{3\times3} : \mathbf{AF} + \mathbf{FA}^\dagger = 0\}. \tag{5.137}$$

Clearly G_β must be a subgroup of G_F. That $G_\beta \neq G_F$ can be seen since the matrix

$$\begin{pmatrix} -1 & 0 & 0 \\ 0 & 1 & 0 \\ 0 & 0 & -1 \end{pmatrix} \tag{5.138}$$

is an element of G_F but not G_β. We can define a conjugation operator ($\hat{\ }$) for any general real 3×3 matrix $\mathbf{N}$ as

$$\widehat{\mathbf{N}} = \mathbf{FN}^T\mathbf{F}^{-1}. \tag{5.139}$$

If $\mathbf{N} \in G_F$ then $\widehat{\mathbf{N}} = \mathbf{N}^{-1}$. The general form of elements

$$\mathbf{A} = \begin{pmatrix} a & b & c \\ d & e & f \\ g & h & j \end{pmatrix}, \tag{5.140}$$

in the Lie algebra must satisfy the condition that $\mathbf{A} = -\widehat{\mathbf{A}}$. This conjugate evaluates to

$$
\widehat{\mathbf{A}} = \begin{pmatrix} 0 & 0 & 1 \\ 0 & -\frac{1}{2} & 0 \\ 1 & 0 & 0 \end{pmatrix} \begin{pmatrix} a & d & g \\ b & e & h \\ c & f & j \end{pmatrix} \begin{pmatrix} 0 & 0 & 1 \\ 0 & -2 & 0 \\ 1 & 0 & 0 \end{pmatrix}
$$

$$
= \begin{pmatrix} c & f & j \\ -\frac{b}{2} & -\frac{e}{2} & -\frac{h}{2} \\ a & d & g \end{pmatrix} \begin{pmatrix} 0 & 0 & 1 \\ 0 & -2 & 0 \\ 1 & 0 & 0 \end{pmatrix}
$$

$$
= \begin{pmatrix} j & -2f & c \\ -\frac{h}{2} & e & -\frac{b}{2} \\ g & -2d & a \end{pmatrix}, \tag{5.141}
$$

and we find that the elements of the Lie algebra have only three free parameters and must be of the form

$$\mathbf{A} = \begin{pmatrix} a & b & 0 \\ d & 0 & \frac{b}{2} \\ 0 & 2d & -a \end{pmatrix}. \tag{5.142}$$

This should be obvious, since the nine elements of the Twiss propagation matrix in Eq. (5.87) only depend on the four elements of the symplectic coordinate transformation matrix in Eq. (5.88). However the determinant of the symplectic matrix is 1, so there are only three free parameters to describe the 2×2 symplectic matrix.

A Cayley transform of the matrix $\mathbf{U}$ can be given by

$$\mathbf{K} = (\mathbf{I} - \mathbf{U})(\mathbf{I} + \mathbf{U})^{-1}, \tag{5.143}$$

with the inverse transform given by

$$\mathbf{U} = (\mathbf{I} + \mathbf{K})^{-1}(\mathbf{I} - \mathbf{K}), \tag{5.144}$$

such that $\mathbf{K}$ has the same symmetry as the elements of the Lie algebra $\mathfrak{g}_\beta$. Another Cayley transform pair is given by

$$\mathbf{L} = \mathbf{F}^{-1}(\mathbf{I} - \mathbf{U})(\mathbf{I} + \mathbf{U})^{-1}, \tag{5.145}$$

$$\mathbf{U} = (\mathbf{I} + \mathbf{FL})^{-1}(\mathbf{I} - \mathbf{FL}), \tag{5.146}$$

where $\mathbf{L}$ is an antisymmetric matrix. These Cayley transform pairs of Eqs. (5.143 to 5.146) are actually more general and hold for any Lie group and algebra pair as defined by G_r and $\mathfrak{g}_r$ in Eqs. (CM: 3.137 and CM: 3.138) with their corresponding $\mathbf{J}$.

For a general 3×3 matrix $\mathbf{W}$ the inverse of $\mathbf{I} - \mathbf{W}$ may be written as

$$(\mathbf{I} - \mathbf{W})^{-1} = \frac{1}{|\mathbf{I} - \mathbf{W}|} \left\{ [1 - \text{tr}(\mathbf{W})]\mathbf{I} + \mathbf{W} + |\mathbf{W}|\,\mathbf{W}^{-1} \right\}, \tag{5.147}$$

or replacing $\mathbf{W}$ by $-\mathbf{W}$, this becomes

$$(\mathbf{I} + \mathbf{W})^{-1} = \frac{1}{|\mathbf{I} + \mathbf{W}|} \left\{ [1 + \text{tr}(\mathbf{W})]\mathbf{I} - \mathbf{W} + |\mathbf{W}|\,\mathbf{W}^{-1} \right\}, \tag{5.148}$$

since $|-\mathbf{W}| = -|\mathbf{W}|$ for an odd order square matrix. Note that for 3×3 matrices, $\mathbf{I} - \mathbf{W}$ commutes with $(\mathbf{I} + \mathbf{W})^{-1}$. (Actually it is easy to show that $\mathbf{I} - \mathbf{W}$ and $(\mathbf{I} + \mathbf{W})^{-1}$ commute for any $n \times n$ matrix $\mathbf{W}$ provided that $|\mathbf{I} + \mathbf{W}| \neq 0$.)

In our case with $\mathbf{W} = \mathbf{U}$, we get

$$\begin{aligned}
\mathbf{K} &= \frac{1}{|\mathbf{I} + \mathbf{U}|} [\mathbf{I} - \text{tr}(\mathbf{U})\mathbf{I} - \mathbf{U} + \mathbf{U}^{-1}](\mathbf{I} - \mathbf{U}) \\
&= \frac{1}{|\mathbf{I} + \mathbf{U}|} [\mathbf{I} - \text{tr}(\mathbf{U})\mathbf{I} - 2\mathbf{U} + \text{tr}(\mathbf{U})\mathbf{U} + \mathbf{U}^2 + \mathbf{U}^{-1} - \mathbf{I}] \\
&= \frac{1}{|\mathbf{I} + \mathbf{U}|} \left\{ [1 - \text{tr}(\mathbf{U})]\mathbf{I} + [\text{tr}(\mathbf{U}) - 2]\mathbf{U} + \mathbf{U}^2 + \mathbf{U}^{-1} \right\}. \tag{5.149}
\end{aligned}$$

When $|\mathbf{U}| = 1$, then the characteristic polynomial $f(x) = |\mathbf{I}x - \mathbf{U}|$ is

$$\begin{aligned}
f(x) &= (x - \lambda_1)(x - \lambda_2)(x - \lambda_3) \\
&= x^3 - \text{tr}(\mathbf{U})x^2 + (\lambda_1\lambda_2 + \lambda_1\lambda_3 + \lambda_2\lambda_3)x - 1. \tag{5.150}
\end{aligned}$$

Recalling the eigenvalues from problem 5-6b, we see that the coefficient of x is just $\text{tr}(\mathbf{U}) = 1 + 2\cos(2\mu)$, so the characteristic polynomial must be

$$f(x) = x^3 - \text{tr}(\mathbf{U})(x^2 - x) - 1, \quad \text{or} \tag{5.151}$$

$$f(x) = (x - 1)\left\{x^2 + [1 - \text{tr}(\mathbf{U})]x + 1\right\}. \tag{5.152}$$

Similarly the characteristic polynomial for $-\mathbf{U}$ will be

$$g(x) = |\mathbf{I}x + \mathbf{U}| = x^3 + \text{tr}(\mathbf{U})(x^2 + x) + 1, \quad \text{or} \tag{5.153}$$

$$g(x) = (x + 1)\left\{x^2 - [1 - \text{tr}(\mathbf{U})]x + 1\right\}. \tag{5.154}$$

Clearly $g(0) = -f(0) = |\mathbf{U}| = 1$, and

$$f(1) = |\mathbf{I} - \mathbf{U}| = 0, \tag{5.155}$$

$$g(1) = |\mathbf{I} + \mathbf{U}| = 2[1 + \text{tr}(\mathbf{U})] = 4[1 + \cos(2\mu)]. \tag{5.156}$$

The Cayley transform fails when the matrix $\mathbf{I} + \mathbf{U}$ does not have an inverse, i. e. when $\mu = \pm\pi/2, \pm3\pi/2, \cdots$. This corresponds to fractional tunes of $\pm 1/4$.

5.9   Problem 5-8: Integral representation of dispersion

a) Show that the particular solution of Eq. (CM: 5.77)

$$\frac{dx}{ds^2} + k(s)x = \frac{1}{\rho(s)}\delta,\qquad (\text{CM}:\ 5.77)$$

may be written as

$$D(s) = S(s)\int_0^s \frac{C(\tau)}{\rho(\tau)}\,d\tau - C(s)\int_0^s \frac{S(\tau)}{\rho(\tau)}\,d\tau.\qquad (5.157)$$

b) Show that the periodic dispersion function may be written as

$$\eta(s) = \frac{\sqrt{\beta(s)}}{2\sin(\pi Q_{\mathrm{H}})}\int_s^{s+L}\frac{\sqrt{\beta(\tau)}}{\rho(\tau)}\cos\left[\phi(\tau) - \phi(s) - \pi Q_{\mathrm{H}}\right]d\tau.\quad (5.158)$$

Note: Perhaps we should have used $\mu(s)$ rather than $\phi(s)$ here to remain consistent with the earlier notation of CM: Chapter 5.

Solution:

a) The functions $C(s)$ and $S(s)$ are solutions of the homogeneous Hill's equation with initial conditions $C(0) = S'(0) = 1$ and $C'(0) = S(0) = 0$, so we have

$$C'' + kC = 0 \quad\text{and}\quad S'' + kS = 0.\qquad (5.159)$$

Likewise, $D(s)$ is the particular solution of the inhomogeneous Hill's equation with $\delta = 1$:

$$D'' + kD = \frac{1}{\rho}.\qquad (5.160)$$

So we need to verify that substitution of Eq. (5.157) into $D'' + kD$ yields $1/\rho$. First we calculate the first and second derivatives of Eq. (5.157):

$$D' = S'\int_0^s \frac{C}{\rho}\,d\tau - \frac{SC}{\rho} - C'\int_0^s \frac{S}{\rho}\,d\tau + \frac{CS}{\rho}$$

$$= S'\int_0^s \frac{C}{\rho}\,d\tau - C'\int_0^s \frac{S}{\rho}\,d\tau,\qquad (5.161)$$

$$D'' = S''\int_0^s \frac{C}{\rho}\,d\tau - C''\int_0^s \frac{S}{\rho}\,d\tau + \frac{CS' - C'S}{\rho}$$

$$= S''\int_0^s \frac{C}{\rho}\,d\tau - C''\int_0^s \frac{S}{\rho}\,d\tau + \frac{1}{\rho}.\qquad (5.162)$$

Substituting into the left-hand side of Eq. (5.160) produces

$$D'' + kD = (S'' + kS) \int_0^s \frac{C}{\rho} d\tau - (C'' + kC) \int_0^s \frac{S}{\rho} d\tau + \frac{1}{\rho} = \frac{1}{\rho},$$
$$(5.163)$$

which is just what we wanted.

b) This solution is similar to the one outlined in Ref. 12. For an alternate solution see § 3.1 of Ref. 59.

Recalling from Eq. (CM: 5.52) we may write

$$C(s|s_0) = \sqrt{\frac{\beta(s)}{\beta(s_0)}} [\cos \mu(s) + \alpha(s_0) \sin \mu(s)], \qquad (5.164)$$

$$S(s|s_0) = \sqrt{\beta(s_0)\beta(s)} \sin \mu(s), \qquad (5.165)$$

$$S'(s|s_0) = \sqrt{\frac{\beta(s_0)}{\beta(s)}} [\cos \mu(s) - \alpha(s) \sin \mu(s)], \qquad (5.166)$$

with

$$\mu(s) = \int_{s_0}^s \frac{d\tau}{\beta(\tau)} = \phi(s) - \phi(s_0). \qquad (5.167)$$

For a full-turn matrix, we have from Eq. (CM: 5.87)[1]

$$\eta(s) = \frac{[1 - S'(s + L|s)]D(s + L|s) + S(s + L|s)D'(s + L|s)}{2(1 - \cos \mu)}, \qquad (5.168)$$

with the one-turn phase advance given by

$$\mu = \phi(s + L) - \phi(s) = \int_s^{s+L} \frac{d\tau}{\beta(\tau)} = 2\pi Q_H. \qquad (5.169)$$

Taking our starting point at $s = s_0$, we may write

$$C(s_0) = S'(s_0) = 1, \quad \text{and} \quad C'(s_0) = S(s_0) = 0. \qquad (5.170)$$

Eqs. (5.157 and 5.161) for the full turn become

$$D(s + L|s) = S(s + L|s) \int_s^{s+L} \frac{C(\tau)}{\rho(\tau)} d\tau - C(s + L|s) \int_s^{s+L} \frac{S(\tau)}{\rho(\tau)} d\tau,$$
$$(5.171)$$

$$D'(s + L|s) = S'(s + L|s) \int_s^{s+L} \frac{C(\tau)}{\rho(\tau)} d\tau - C'(s + L|s) \int_s^{s+L} \frac{S(\tau)}{\rho(\tau)} d\tau.$$
$$(5.172)$$

---

[1]The double argument with a vertical bar "$(s_j|s_i)$" is to be interpreted as a matrix or transport function evaluated from s_i to s_j. Here we have written Eq. (CM: 5.87) using these double arguments to emphasize the 1-turn elements, even though C, S, C', and S' are still all periodic functions.

The denominator of the right-hand side of Eq. (5.168) may be written in terms of the tune as

$$2(1 - \cos\mu) = 4\sin^2(\pi Q_H), \tag{5.173}$$

and the numerator as

$$[1 - S'(s+L|s)]D(s+L|s) + S(s+L|s)D'(s+L|s)$$

$$= [1 - S'(s+L|s)]\left[S(s+L|s) \int_s^{s+L} \frac{C(\tau)}{\rho(\tau)}\,d\tau \right.$$

$$\left. -C(s+L|s) \int_s^{s+L} \frac{S(\tau)}{\rho(\tau)}\,d\tau \right]$$

$$+ S(s+L|s)\left[S'(s+L|s) \int_s^{s+L} \frac{C(\tau)}{\rho(\tau)}\,d\tau \right.$$

$$\left. -C'(s+L|s) \int_s^{s+L} \frac{S(\tau)}{\rho(\tau)}\,d\tau \right]$$

$$= \int_s^{s+L} \frac{S(s+L|s)C(\tau) - C(s+L|s)S(\tau)}{\rho(\tau)}\,d\tau$$

$$+ \int_s^{s+L} \frac{S(\tau)}{\rho(\tau)}\,d\tau. \tag{5.174}$$

Remembering that $\beta(s)$ and $\alpha(s)$ are periodic, Eqs. (5.164 and 5.165) may be combined to yield

$$S(s+L|s)C(\tau) = \sqrt{\beta(\tau)\beta(s)}\,[\cos\mu(\tau)\sin\mu(s+L)$$

$$+ \alpha(s)\sin\mu(\tau)\sin\mu(s+L)], \quad (5.175)$$

$$C(s+L|s)S(\tau) = \sqrt{\beta(s)\beta(\tau)}\,[\cos\mu(s+L)\sin\mu(\tau)$$

$$+ \alpha(s)\sin\mu(s+L)\sin\mu(\tau)], \quad (5.176)$$

whose difference is

$$S(s+L|s)C(\tau) - C(s+L|s)S(\tau) = \sqrt{\beta(s)\beta(\tau)}\,\sin[\mu(s+L) - \mu(\tau)]$$

$$= \sqrt{\beta(s)\beta(\tau)}\,\sin[\mu - \mu(\tau)], \tag{5.177}$$

since $\mu(s+L) = \mu$.

Combining Eqs. (5.168, 5.169, 5.173, 5.174 and 5.177) we may write the dispersion function as

$$\eta(s) = \frac{\sqrt{\beta(s)}}{4\sin^2(\pi Q_H)} \int_s^{s+L} \frac{\sqrt{\beta(\tau)}}{\rho(\tau)}\,\{\sin[\mu - \mu(\tau)] + \sin\mu(\tau)\}\,d\tau. \tag{5.178}$$

The expression inside the braces may be rearranged to

$$\{\cdots\} = \sin\mu \, \cos\mu(\tau) - \cos\mu \, \sin\mu(\tau) + \sin\mu(\tau)$$
$$= 2\sin\frac{\mu}{2} \, \cos\frac{\mu}{2} \, \cos\mu(\tau) + 2\sin^2\frac{\mu}{2} \, \sin\mu(\tau)$$
$$= 2\sin(\pi Q_{\mathrm{H}}) \, \cos\left[\phi(\tau) - \phi(s) - \pi Q_{\mathrm{H}}\right]. \tag{5.179}$$

Substituting back into Eq. (5.178) we finally arrive at the desired

$$\eta(s) = \frac{\sqrt{\beta(s)}}{2\sin(\pi Q_{\mathrm{H}})} \int_s^{s+L} \frac{\sqrt{\beta(\tau)}}{\rho(\tau)} \cos[\phi(\tau) - \phi(s) - \pi Q_{\mathrm{H}}] \, d\tau. \tag{5.180}$$

5.10   Problem 5-9: Betatron phase advance

Given the transfer matrix for a series of elements and the initial Twiss parameters, what is the betatron phase advance through this set of elements?

Solution:

Recalling Eq. (CM:5.52) the transfer matrix may be written as

$$\mathbf{M} = \begin{pmatrix} C & S \\ C' & S' \end{pmatrix},$$

(5.181)

where

$$C = \sqrt{\frac{\beta}{\beta_0}}[\cos\mu + \alpha_0 \sin\mu],$$

(5.182)

$$S = \sqrt{\beta\beta_0}\sin\mu,$$

(5.183)

$$C' = -\frac{(\alpha - \alpha_0)\cos\mu + (1 + \alpha\alpha_0)\sin\mu}{\sqrt{\beta\beta_0}},$$

(5.184)

$$S' = \sqrt{\frac{\beta_0}{\beta}}[\cos\mu - \alpha\sin\mu].$$

(5.185)

From problem 5–6, we may write

$$\beta = \beta_0 C^2 - 2\alpha_0 CS + \gamma_0 S^2,$$

$$\alpha = -\beta_0 CC' + \alpha_0 (CS' + C'S) - \gamma_0 SS'.$$

(5.186)

Solving Eq. (5.183) for $\sin\mu$ and Eq. (5.182) for $\cos\mu$ gives

$$\sin\mu = \frac{S}{\sqrt{\beta\beta_0}} = \frac{S}{\sqrt{\beta_0^2 C^2 - 2\alpha_0\beta_0 CS + \beta_0\gamma_0 S^2}}$$

$$= \frac{S}{\sqrt{\beta_0^2 C^2 - 2\alpha_0\beta_0 CS + (1 + \alpha_0^2)S^2}}$$

$$= \frac{S}{\sqrt{S^2 + (\beta_0 C - \alpha_0 S)^2}},$$

(5.187)

$$\cos\mu = C\frac{\beta_0}{\beta} - \alpha_0 \sin\mu = \frac{\beta_0 C - \alpha_0 S}{\sqrt{S^2 + (\beta_0 C - \alpha_0 S)^2}}.$$

$$\tan\mu = \frac{S}{\beta_0 C - \alpha_0 S}.$$

(5.188)

$$\mu = \tan^{-1}\left(\frac{S}{\beta_0 C - \alpha_0 S}\right).$$

(5.189)

If $\cos\mu < 0$, then add π to μ. Of course, we must also realize the phase advance determined in this way is actually just μ mod 2π.

5.11 Problem 5-10: Conversion of emittances

Show that the conversion from rms to 90% and 95% emittances are approximately $\epsilon_{90\%} \simeq 4.605\epsilon_{\text{rms}}$, and $\epsilon_{95\%} \simeq 5.991\epsilon_{\text{rms}}$, for a Gaussian distribution. Note: Here the second $\epsilon_{90\%}$ has been corrected to read $\epsilon_{95\%}$.

Solution:

From Eq. (5.48) of Problem 5–4, we have the distribution

$$f(\xi, \zeta) = \frac{N}{2\pi\epsilon_{\text{rms}}} \exp\left(-\frac{\xi^2 + \zeta^2}{2\epsilon_{\text{rms}}}\right), \tag{5.190}$$

where ϵ_{rms} is the rms emittance, and ξ and ζ are the rescaled coordinates to turn the ellipse into a circle. This is a 2-d Gaussian distribution with an rms size $\sigma = \sqrt{\epsilon_{\text{rms}}}$. Setting $N = 1$ we can use this as a probability distribution. Integrating out to a radius R gives us the fraction of particles inside a circle of radius R:

$$
\begin{aligned}
F(R) &= \frac{1}{2\pi\epsilon_{\text{rms}}} \int_{-R}^{R} \int_{-\sqrt{R^2-x^2}}^{\sqrt{R^2-\xi^2}} \exp\left(-\frac{\xi^2 + \zeta^2}{2\epsilon_{\text{rms}}}\right) d\xi \, d\zeta \\
&= \frac{1}{2\pi\epsilon_{\text{rms}}} \int_{0}^{R} \int_{0}^{2\pi} \exp\left(-\frac{r^2}{2\epsilon_{\text{rms}}}\right) r \, d\theta \, dr \\
&= 1 - e^{-R^2/(2\epsilon_{\text{rms}})}.
\end{aligned}
\tag{5.191}
$$

For $R = 1\sigma$ we find that $F(1\sigma) = 0.393$ of the beam is within the 1-σ contour. Solving for $(R/\sigma)^2$ we have

$$\epsilon_F/\epsilon_{\text{rms}} = R^2/\sigma^2 = -2\ln(1 - F) \tag{5.192}$$

with some values given in Table 5.1.

Table 5.1   Fraction of beam

$F\left(\sqrt{\frac{\epsilon_F}{\epsilon_{\text{rms}}}}\sigma\right)$	$\epsilon_F/\epsilon_{\text{rms}}$
39.35%	1.000
50%	1.386
75%	2.773
85%	3.794
90%	4.605
95%	5.991
98%	7.824

5.12   Problem 5-11: Emittance growth in foils

A gold (^{197}Au^{+79}) beam passes through a 1 mm thick Al$_2$O$_3$ flag, at a location in the beam line where the Twiss parameters are $\beta_H = \beta_V = 6$ m and $\alpha_H = \alpha_V = 0$. Multiple Coulomb scattering takes place in the flag adding to the angular divergence of the beam. A good approximation[42] of the deflection is a Gaussian distribution with an rms angle given by

$$\bar{\theta} = \sqrt{\langle\theta^2\rangle} \simeq z\frac{20 \text{ MeV/c}}{p\beta}\sqrt{\frac{x}{L_{\text{rad}}}}\left(1 + \frac{1}{9}\log_{10}\frac{x}{L_{\text{rad}}}\right), \qquad (5.193)$$

where z is the beam particle's charge, x is the thickness of the flag expressed in [g $\cdot$ cm^{-2}]. Assume[66] $L_{\text{rad}} = 24$ g $\cdot$ cm^{-2} and $\rho_{\text{Al}_2\text{O}_3} = 3.7$ g $\cdot$ cm^{-3}. The gold beam has a total energy of 10 GeV/nucleon with $M_{\text{Au}}/A = 0.93113$ GeV/nucleon.

a) Evaluate $\bar{\theta}$.

b) The beam has a normalized emittance $\epsilon^N_{95\%} = 10$ μm just before the flag. Estimate the blowup in emittance from the flag.

Note: We should have mentioned that the $\bar{\theta}$ in the above equation is calculated for an opening cone angle. What we require for a single plane[42] must be projected into the plane

$$\bar{\theta}_x = \frac{\bar{\theta}}{\sqrt{2}}. \qquad (5.194)$$

Solution:

a)

$$\gamma = 1 + \frac{K}{mc^2} = 1 + \frac{179 \times 10 \text{ GeV}}{179 \times 0.93313 \text{ GeV}} = 11.74, \qquad (5.195)$$

$$p\beta c = \beta^2\gamma mc^2 = \frac{\gamma^2 - 1}{\gamma}mc^2 = 1943 \text{ GeV}, \qquad (5.196)$$

$$\frac{x}{L_{\text{rad}}} = \frac{\rho_{\text{Al}_2\text{O}_3}t}{L_{\text{rad}}} = \frac{3.7 \times 0.1}{24} = 0.0154, \qquad (5.197)$$

$$\bar{\theta} = 79 \times \frac{0.02 \text{ GeV}}{1943 \text{ GeV}}\sqrt{0.0154}\left(1 + \frac{\log_{10}(0.0154)}{9}\right)$$

$$= 81 \text{ }\mu\text{rad},$$

$$\bar{\theta}_x = 57 \text{ }\mu\text{rad}. \qquad (5.198)$$

b) Using a result from the previous problem

$$\epsilon_{\text{rms}} = \frac{\epsilon_{95\%}^N}{5.991\beta\gamma} = 0.143 \ \mu\text{m}, \tag{5.199}$$

where $\beta\gamma = \sqrt{\gamma^2 - 1} = 11.70$. The rms size of the beam at the foil is then $\sigma = \sqrt{\beta_{\text{H}} \, \epsilon_{\text{rms}}} \simeq 0.925$ mm. The shift of the position due to multiple scattering going through the foil should be less than

$$\Delta x \lesssim \bar{\theta}_x t \sim 47 \ \text{nm}. \tag{5.200}$$

The rms divergence at the upstream edge of the foil is

$$\sigma' = \sqrt{\gamma_{\text{H}} \, \epsilon_{\text{rms}}} = \frac{\sigma}{\beta_{\text{H}}} = 154 \ \mu\text{rad}. \tag{5.201}$$

Comparing the effect of multiple scattering on both the x and x' dimensions of phase space we find $\bar{\theta}_x/\sigma' \sim 0.4$ whereas $\Delta x/\sigma \sim 10^{-4}$, so we can ignore contribution from the spatial dimensions and just consider the angular growth.

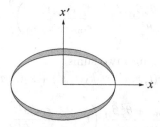

Fig. 5.2 Multiple scattering increases the divergence (x'-axis) as the beam passes through the flag, but the spatial contribution is miniscule. The shaded region show the effective increase of a 1σ contour. The ellipses are plotted for $\alpha = 0$.

At the upstream edge of the flag, a Gaussian beam distribution in phase space is given by

$$b_{\text{up}}(x, x') = \frac{N}{2\pi\sigma\sigma'} \exp\left[-\left(\frac{x^2}{2\beta_{\text{H}}\epsilon_{\text{rms}}} + \frac{x'^2}{2\epsilon_{\text{rms}}\gamma_{\text{H}}} \right) \right]$$

$$= \frac{N}{2\pi\epsilon_{\text{rms}}} \exp\left[-\left(\frac{\gamma_{\text{H}}x^2 + \beta_{\text{H}}x'^2}{2\epsilon_{\text{rms}}} \right) \right] \tag{5.202}$$

where $\pi\epsilon_{\text{rms}}$ is the unnormalized rms emittance and N is the number of particles in the beam. In the small angle approximation, the distribution of scattered angles is

$$g_{\bar{\theta}_x}(x') = \frac{1}{\sqrt{2\pi}\,\bar{\theta}_x} e^{-x'^2/(2\bar{\theta}_x^2)}. \tag{5.203}$$

The multiple scattering will smear the distribution at the downstream edge of the flag (see Fig. 5.2) and can be calculated by the convolution:

$$b_{\mathrm{dn}}(x, x') \int_{-\infty}^{\infty} g_{\bar{\theta}_x}(x' - t) b_{\mathrm{up}}(x, t)\, dt$$

$$= \frac{N}{(2\pi)^{3/2}\epsilon_{\mathrm{rms}}\bar{\theta}_x} \int_{-\infty}^{\infty} e^{-k(t,x,x')}\, dt, \qquad (5.204)$$

where

$$
\begin{aligned}
k(t, x, x') &= \frac{(x' - t)^2}{2\bar{\theta}_x^2} + \frac{\gamma_{\mathrm{H}} x^2 + \beta_{\mathrm{H}} t^2}{2\epsilon_{\mathrm{rms}}} \\
&= \frac{\epsilon_{\mathrm{rms}} x'^2 + \bar{\theta}_x^2 \gamma_{\mathrm{H}} x^2}{2\epsilon_{\mathrm{rms}} \bar{\theta}_x^2} + \frac{(\epsilon_{\mathrm{rms}} + \bar{\theta}_x^2 \beta_{\mathrm{H}}) t^2 - 2\epsilon_{\mathrm{rms}} x' t}{2\epsilon_{\mathrm{rms}} \bar{\theta}_x^2} \\
&= \frac{\epsilon_{\mathrm{rms}} x'^2 + \bar{\theta}_x^2 \gamma_{\mathrm{H}} x^2}{2\epsilon_{\mathrm{rms}} \bar{\theta}_x^2} + \frac{(\epsilon_{\mathrm{rms}} + \bar{\theta}_x^2 \beta_{\mathrm{H}}) t^2 - 2\epsilon_{\mathrm{rms}} x' t}{2\epsilon_{\mathrm{rms}} \bar{\theta}_x^2}. \quad (5.205)
\end{aligned}
$$

By completing the square in the second term, we have

$$
\begin{aligned}
\frac{(\epsilon_{\mathrm{rms}} + \bar{\theta}_x^2 \beta_{\mathrm{H}}) t^2 - 2\epsilon_{\mathrm{rms}} x' t}{2\epsilon_{\mathrm{rms}} \bar{\theta}_x^2} &= \frac{\epsilon_{\mathrm{rms}} + \bar{\theta}_x^2 \beta_{\mathrm{H}}}{2\epsilon_{\mathrm{rms}} \bar{\theta}_x^2} \left(t^2 - \frac{2\epsilon_{\mathrm{rms}} x' t}{\epsilon_{\mathrm{rms}} + \bar{\theta}_x^2 \beta_{\mathrm{H}}} \right) \\
&= \frac{\epsilon_{\mathrm{rms}} + \bar{\theta}_x^2 \beta_{\mathrm{H}}}{2\epsilon_{\mathrm{rms}} \bar{\theta}_x^2} \left(t - \frac{\epsilon_{\mathrm{rms}} x'}{\epsilon_{\mathrm{rms}} + \bar{\theta}_x^2 \beta_{\mathrm{H}}} \right)^2 - \frac{\epsilon_{\mathrm{rms}} x'^2}{2\bar{\theta}_x^2 (\epsilon_{\mathrm{rms}} + \bar{\theta}_x^2 \beta_{\mathrm{H}})}
\end{aligned}
$$
$$(5.206)$$

So $k(t, x, x')$ separates into two sums

$$k(t, x, x') = k_1(t, x') + k_2(x, x'), \qquad (5.207)$$

with

$$k_1(t, x') = \frac{\epsilon_{\mathrm{rms}} + \bar{\theta}_x^2 \beta_{\mathrm{H}}}{2\epsilon_{\mathrm{rms}} \bar{\theta}_x^2} \left(t - \frac{\epsilon_{\mathrm{rms}} x'}{\epsilon_{\mathrm{rms}} + \bar{\theta}_x^2 \beta_{\mathrm{H}}} \right)^2, \qquad (5.208)$$

$$k_2(x, x') = \frac{\epsilon_{\mathrm{rms}} x'^2 + \bar{\theta}_x^2 \gamma_{\mathrm{H}} x^2}{2\epsilon_{\mathrm{rms}} \bar{\theta}_x^2} - \frac{\epsilon_{\mathrm{rms}} x'^2}{2\bar{\theta}_x^2 (\epsilon_{\mathrm{rms}} + \bar{\theta}_x^2 \beta_{\mathrm{H}})}, \qquad (5.209)$$

and

$$
\begin{aligned}
b_{\mathrm{dn}}(x, x') &= \frac{N\, e^{-k_2(x,x')}}{(2\pi)^{3/2}\epsilon_{\mathrm{rms}}\bar{\theta}_x} \int_{\infty}^{\infty} e^{-k_1(t,x')}\, dt \\
&= \frac{N\, e^{-k_2(x,x')}}{2\pi \sqrt{\epsilon_{\mathrm{rms}}(\epsilon_{\mathrm{rms}} + \bar{\theta}_x^2 \beta_{\mathrm{H}})}} \\
&= \frac{N\, e^{-k_2(x,x')}}{2\pi \epsilon_{\mathrm{rms}} \sqrt{1 + \frac{\beta_{\mathrm{H}} \bar{\theta}_x^2}{\epsilon_{\mathrm{rms}}}}}. \qquad (5.210)
\end{aligned}
$$

Comparing with Eq. (5.202) we see that the emittance increases by the ratio

$$\frac{\epsilon_{\mathrm{dn}}}{\epsilon_{\mathrm{up}}} = \sqrt{1 + \frac{\beta_{\mathrm{H}} \bar{\theta}_x^2}{\epsilon_{\mathrm{rms}}}} = \sqrt{1 + \frac{6 \times (5.7 \times 10^{-5})^2}{1.43 \times 10^{-7}}} \simeq 1.07. \qquad (5.211)$$

Comments:

We should also note that (see Fig. 5.2) the shape of the ellipse has changed which implies that the Courant-Snyder parameters have been changed. Rearranging $k_2(x, x')$ gives

$$k_2(x, x') = \frac{\dfrac{\beta_H \epsilon_{\mathrm{rms}}}{\sqrt{\epsilon_{\mathrm{rms}}(\epsilon_{\mathrm{rms}} + \bar{\theta}_x^2 \beta_H)}} x'^2 + \dfrac{\gamma_H \epsilon_{\mathrm{rms}} + \bar{\theta}_x^2}{\sqrt{\epsilon_{\mathrm{rms}}(\epsilon_{\mathrm{rms}} + \bar{\theta}_x^2 \beta_H)}} x^2}{2\sqrt{\epsilon_{\mathrm{rms}}(\epsilon_{\mathrm{rms}} + \bar{\theta}_x^2 \beta_H)}}. \tag{5.212}$$

So we may write

$$\epsilon_{\mathrm{rms}}^* = \epsilon_{\mathrm{rms}} \sqrt{1 + \frac{\bar{\theta}_x^2 \beta_H}{\epsilon_{\mathrm{rms}}}}, \tag{5.213}$$

$$\beta_H{}^* = \frac{\beta_H}{\sqrt{1 + \frac{\bar{\theta}_x^2 \beta_H}{\epsilon_{\mathrm{rms}}}}}, \tag{5.214}$$

$$\gamma_H{}^* = \frac{\gamma_H + \frac{\bar{\theta}_x^2}{\epsilon_{\mathrm{rms}}}}{\sqrt{1 + \frac{\bar{\theta}_x^2 \beta_H}{\epsilon_{\mathrm{rms}}}}}, \tag{5.215}$$

with the superscript asterisk indicating values at the downstream edge of the flag. Note that $\sigma = \sqrt{\epsilon_{\mathrm{rms}}\beta_H} = \sqrt{\epsilon_{\mathrm{rms}}^* \beta_H{}^*} = \sigma^*$ as we should expect since we neglected the small Δx contribution.

We leave as an exercise for the reader to show that in the case of $\alpha \neq 0$, then these formulae still hold with the additional formula

$$\alpha_H{}^* = \frac{\alpha_H}{\sqrt{1 + \frac{\bar{\theta}_x^2 \beta_H}{\epsilon_{\mathrm{rms}}}}}. \tag{5.216}$$

Chapter 6

Problems of Chapter 6: Lattice Exercises

6.1 Problem 6–1: Maximum phase of drift and FODO cell

a) What is the maximum possible phase advance in a drift?
b) What is the maximum possible phase advance in a FODO cell?

Solution:

a) Recall (see CM: § 5.2) that we may write the phase advance between two azimuths as

$$\Delta\Psi = \Psi(s_1) - \Psi(s_0) = \int_{s_0}^{s_1} \frac{ds}{\beta(s)}. \tag{6.1}$$

Using the result of Problem 5–5, the evolution of a beta-function in a drift may be written as

$$\beta(s) = \beta_0 - 2\alpha_0 s + \gamma_0 s^2. \tag{6.2}$$

Note that this is the equation of a parabola. Clearly the largest value we can get for $\Delta\Psi$ will be if we put the minimum of $\beta^* = \beta(s_{\min})$ in the middle of the drift. For simplicity let us take $s_{\min} = 0$ so that

$$\beta(s) = \beta_* + \frac{s^2}{\beta_*}. \tag{6.3}$$

The indefinite integral in Eq. (6.1) evaluates as

$$\int \frac{ds}{\beta_* + \frac{s^2}{\beta_*}} = \int \frac{d(s/\beta^*)}{1 + (s/\beta^*)^2} = \tan^{-1}\left(\frac{s}{\beta^*}\right). \tag{6.4}$$

If we take the limit of an infinitely long straight section with $s_0 = -\infty$ and $s_1 = +\infty$, then we find

$$\Delta\Psi = 90° - (-90°) = 180° \tag{6.5}$$

as the maximum limit for the phase advance in a drift.

b) For a FODO cell we would like to have stability in both planes, so in a given plane we must have one quadrupole (the "F" quad) at β_{max} and the other (the "D" quad) at β_{min}. In the other plane, the locations of β_{min} and β_{max} will be reversed. Therefore the limiting value of phase advance for stable motion in half of a FODO cell must be 90°. Adding the two halves together gives us again a limit of 180° for the maximum total phase advance in a FODO cell.

Comment:

In practice, more common values used for phase advances in FODO cells in most lattices tend to be in the range 60° to 120°. Values near 180° give large β_{max} requiring large apertures. Small values near 0° would give a FODO lattice with probably more quadrupoles than necessary, leading to a more expensive lattice. (A counter example of this is the weak focusing and azimuthally symmetric "betatron" lattice of CM: Chapter 2, if we were to consider this as a FODO lattice with an infinite number of zero-length FODO cells.)

6.2 Problem 6–2: Design of a FODO lattice

Design a lattice with four identical FODO cells. Let the basic cell consist of a thin lens, a sector magnet, another thin lens, and another sector magnet. Pick suitable values for the two quadrupole strengths, drift length, and bending radius, so that there is stability in both dimensions.

a) What are the horizontal and vertical betatron tunes?
b) Plot $\beta_x(s)$, $\beta_y(s)$, and $\eta(s)$ for the periodic cell.
c) Calculate the momentum compaction.

Solution:

Let us pick a unit cell consisting as follows:

$$M_{cell} = \textbf{FLBLDLBL}, \qquad (6.6)$$

where the matrices are as follows:

$$\textbf{F} : \text{a thin horizontally focusing quad of strength } k_f, \qquad (6.7)$$
$$\textbf{D} : \text{a thin horizontally defocusing quad of strength } k_d, \qquad (6.8)$$
$$\textbf{L} : \text{a short drift of length } l, \qquad (6.9)$$
$$\textbf{B} : \text{a } 360°/8 = 45° \text{ bend of radius } \rho. \qquad (6.10)$$

For simplicity let us use values the values: $l = 1$ m and $\rho = 10$ m. the arc length of the design orbit through a bend must then be $l_{bend} = \pi\rho/4 = 7.854$ m, so the total length of the cell must be

$$l_{cell} = 2 \times l_{bend} + 4 \times l = 19.708 \text{ m}, \qquad (6.11)$$

which gives a total circumference of 78.832 m.

In order to calculate the tunes and lattice functions, we have written a program in the Octave[24] computer language. Functions producing matrices for transport orbits and lattice functions are given below in §6.2.1, and an Octave program to calculate the data is listed in §6.2.2.

a) The Octave program calculates tunes:

$$Q_H = 1.157599, \quad \text{and} \quad Q_V = 0.460107, \qquad (6.12)$$

with the focal lengths of 19.708 m and -16.423 m respectively for the horizontally focusing and defocusing quadrupoles.

b) The program uses the methods of CM: Chapter 5 to calculate the lattice parameters as a function of s. We have sliced the drifts and bends each into 10 pieces, so that the curves would appear smooth. Propagation of the Twiss parameters were performed using the matrix formalism of Problem 5–5 (see § 5.5) and the results are plotted in Fig. 6.1.

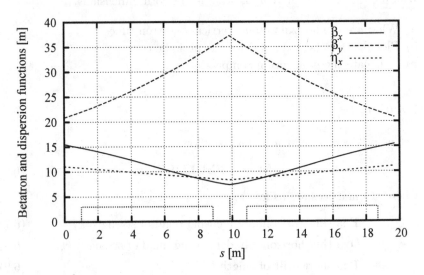

Fig. 6.1 Lattice functions $\beta_x(s)$, $\beta_y(s)$, and $\eta_x(s)$ plotted for the periodic FODO cell.

c) The momentum compaction can be calculated by the integral

$$\alpha_p = \frac{1}{L} \oint \frac{\eta_x}{\rho} \, ds$$

$$= \frac{8}{L} \int_0^{\pi/4} \eta_x(\theta) \, d\theta,$$

(6.13)

since the contribution comes only from the bends, and η_x is mirrored about the quadrupole centers with the symmetry we chose for our lattice. The dispersion function in a bend propagates as

$$\eta_x(\theta) = \eta_0 \cos \theta + \eta_0' \rho \sin \theta + \rho(1 - \cos \theta),$$ (6.14)

with initial values

$$\begin{pmatrix} \eta_0 \\ \eta_0' \end{pmatrix} = \begin{pmatrix} 10.773812 \text{ m} \\ -0.280452 \end{pmatrix}$$ (6.15)

at the upstream end of the first dipole. Performing the integration we have

$$\alpha_p = \frac{8}{L} \left[\eta_0 \sin\theta - \eta_0'\rho \cos\theta + \rho(\theta - \sin\theta) \right] \Big|_0^{\pi/4} = 0.7692, \qquad (6.16)$$

which is fairly close to $Q_{\mathrm{H}}^{-2} = 0.746$.

6.2.1 *Listing of* Octave *functions for transport matrices*

Function for a drift matrix with filename: "drift.m"

```
function A = drift (x, gam) % drift matrix
A = zeros(6,6);
A=[1,x,0,0,0,0; 0,1,0,0,0,0;        0,0,1,x,0,0;
   0,0,0,1,0,0; 0,0,0,0,1,x/gam^2; 0,0,0,0,0,1];
end
```

Function for a quadrupole matrix with filename: "quadrupole.m"

```
function A = quadrupole (kk, len, gam) % matrix for a quad
k = sqrt(kk);
a = k*len;
cf = cos(a);
sf = sin(a);
cd = cosh(a);
sd = sinh(a);
A = zeros(6,6);
A=[cf,     sf/k,    0,    0,    0,          0;
   -k*sf,  cf,      0,    0,    0,          0;
   0,      0,       cd,   sd/k, 0,          0;
   0,      0,       k*sd, cd,   0,          0;
   0,      0,       0,    0,    1.0,  len/gam^2;
   0,      0,       0,    0,    0,          1.0];
end
```

Function for a sector bend matrix with filename: "sbend.m"

```
function A = sbend (rho, theta, gam)
  % matrix for a sector bend
A = zeros(6,6);
A = [cos(theta), rho*sin(theta), 0, 0, 0, rho*(1.-cos(theta));
     -sin(theta)/rho, cos(theta), 0, 0, 0, sin(theta);
```

```
    0, 0, 1.0, rho*theta, 0,  0;
    0, 0, 0,  1.0,  0,  0;
    -sin(theta), -rho*(1.-cos(theta)), 0, 0, 1, \
        rho*(-theta+sin(theta)+theta/gam^2);
    0, 0, 0,  0,  0,  1;
   ];
end
```

Function for a the Twiss matrix (see §5.5) filename: "Twiss.m"

```
function M=Twiss(T)
  M(1,1)=T(1,1)**2;
  M(1,2)=-2*T(1,1)*T(1,2);
  M(1,3)=T(1,2)**2;
  M(2,1)=-T(1,1)*T(2,1);
  M(2,2)=T(1,1)*T(2,2)+T(1,2)*T(2,1);
  M(2,3)=-T(1,2)*T(2,2);
  M(3,1)=T(2,1)**2;
  M(3,2)=-2*T(2,1)*T(2,2);
  M(3,3)=T(2,2)**2;
endfunction
```

6.2.2 *Listing of the* Octave *program*

```
#!/usr/bin/octave --persist
%#!/usr/bin/octave
% Conte and MacKay problem 6--2. Simple FODO lattice with
% periodicity 4.

% Pick a Lorentz factor for the particle (say a proton)
mc2 = 0.938272;      % GeV
U = 3.0;             % GeV
gam = U/mc2;

ldrift = 1.0;
theta = pi/4.0;
rho = 10.;           % m
lbend = rho*theta;

L = drift(ldrift, gam);
```

```
B = sbend(rho, theta, gam);

lq=1.e-6; % m (Approx thin quad with thick lens.)

lcell = 2*lbend+4*ldrift+2*lq;
% Make a stab at a stable focal length:

kf =   1.0/lcell/lq;
kd = -1.2/lcell/lq;
F = quadrupole(kf, lq, gam);
D = quadrupole(kd, lq, gam);

Mcell = L*B*L*D*L*B*L*F;

% Check stability:
if( abs(Mcell(1,1)+Mcell(2,2))>=2. )
  printf("Mcell is horizontally unstable.\n");
endif;
if( abs(Mcell(3,3)+Mcell(4,4))>=2. )
  printf("Mcell is vertically unstable.\n");
endif;

%%%%%%%%%%%%%%%%%%%%%% Part a  %%%%%%%%%%%%%%%%%%%%
% Part a    Tunes calculated from trace of 2x2 blocks:
mucellh = acos( (Mcell(1,1)+Mcell(2,2))/2. );
mucellv = acos( (Mcell(3,3)+Mcell(4,4))/2. );
Qh = 4.0 * mucellh/2./pi;
Qv = 4.0 * mucellv/2./pi;
printf("Problem 6-2, Part a:   Qh=%f\t Qv=%f\n", Qh, Qv);

%%%%%%%%%%%%%%%%%%%%%% Part b  %%%%%%%%%%%%%%%%%%%%
% Get the starting values of the betax, alphax, betay, alphay
% and etax:
betax0 = Mcell(1,2)/sin(mucellh);
alphax0 = (Mcell(1,1)-Mcell(2,2))/2./sin(mucellh);
betay0 = Mcell(3,4)/sin(mucellv);
alphay0 = (Mcell(3,3)-Mcell(4,4))/2./sin(mucellv);
etax0 = ((1.-Mcell(2,2))*Mcell(1,6)+Mcell(1,2)*Mcell(2,6)) \
   /2./(1.-cos(mucellh));
```

```
etaxp0 = ((1.-Mcell(1,1))*Mcell(2,6)+Mcell(2,1)*Mcell(1,6)) \
   /2./(1.-cos(mucellh));
s = 0

% Calculate and write out the optics functions for one cell.
nsteps = 10.;       % Subdivide drifts and bends.
dd = drift(ldrift/nsteps, gam);
bb = sbend(rho, theta/nsteps, gam);

fid = fopen("optic.dat", "w")  % save in this file.
fprintf(fid, "#  s[m],        betax [m],  betay [m],");
fprintf(fid, "     etax [m]    elt\n");
s = 0;
bagx = bagx0 = [betax0; alphax0; (1.+alphax0**2)/betax0];
bagy = bagy0 = [betay0; alphay0; (1.+alphay0**2)/betay0];
dsp  = dsp0  = [etax0; etaxp0; 0; 0; 0; 1.];
fprintf(fid, "%10.6f %10.3f %10.3f %10.3f    start\n", \
        s, bagx(1), bagy(1), dsp(1));
% propagate through first quad
s = s+lq;
dsp = F*dsp;
bagx = Twiss(F(1:2,1:2))*bagx;
bagy = Twiss(F(3:4,3:4))*bagy;
fprintf(fid, "%10.6f %10.3f %10.3f %10.3f    q\n", \
        s, bagx(1), bagy(1), dsp(1));

% propagate through 1st drift
for s = s+ldrift/nsteps:ldrift/nsteps:s+ldrift
  dsp = dd*dsp;
  bagx = Twiss(dd(1:2,1:2))*bagx;
  bagy = Twiss(dd(3:4,3:4))*bagy;
  fprintf(fid, "%10.6f %10.3f %10.3f %10.3f    d\n", \
      s, bagx(1), bagy(1), dsp(1));
endfor;
% propagate through bend
for s = s+lbend/nsteps:lbend/nsteps:s+lbend
  dsp = bb*dsp;
  bagx = Twiss(bb(1:2,1:2))*bagx;
  bagy = Twiss(bb(3:4,3:4))*bagy;
```

```
  fprintf(fid, "%10.6f  %10.3f  %10.3f  %10.3f    b\n", \
      s, bagx(1), bagy(1), dsp(1));
endfor;
% propagate through 2nd drift
for s = s+ldrift/nsteps:ldrift/nsteps:s+ldrift
  dsp = dd*dsp;
  bagx = Twiss(dd(1:2,1:2))*bagx;
  bagy = Twiss(dd(3:4,3:4))*bagy;
  fprintf(fid, "%10.6f  %10.3f  %10.3f  %10.3f    d\n", \
      s, bagx(1), bagy(1), dsp(1));
endfor;
% propagate through 2nd quad
s = s+lq;
dsp = D*dsp;
bagx = Twiss(D(1:2,1:2))*bagx;
bagy = Twiss(D(3:4,3:4))*bagy;
fprintf(fid, "%10.6f  %10.3f  %10.3f  %10.3f    q\n", \
      s, bagx(1), bagy(1), dsp(1));
% propagate through 3rd drift
for s = s+ldrift/nsteps:ldrift/nsteps:s+ldrift
  dsp = dd*dsp;
  bagx = Twiss(dd(1:2,1:2))*bagx;
  bagy = Twiss(dd(3:4,3:4))*bagy;
  fprintf(fid, "%10.6f  %10.3f  %10.3f  %10.3f    d\n", \
      s, bagx(1), bagy(1), dsp(1));
endfor;
% propagate through bend
for s = s+lbend/nsteps:lbend/nsteps:s+lbend
  dsp = bb*dsp;
  bagx = Twiss(bb(1:2,1:2))*bagx;
  bagy = Twiss(bb(3:4,3:4))*bagy;
  fprintf(fid, "%10.6f  %10.3f  %10.3f  %10.3f    b\n", \
      s, bagx(1), bagy(1), dsp(1));
endfor;
% propagate through 4th drift
for s = s+ldrift/nsteps:ldrift/nsteps:s+ldrift
  dsp = dd*dsp;
  bagx = Twiss(dd(1:2,1:2))*bagx;
  bagy = Twiss(dd(3:4,3:4))*bagy;
```

```
    fprintf(fid, "%10.6f   %10.3f   %10.3f   %10.3f    d\n", \
        s, bagx(1), bagy(1), dsp(1));
endfor;
fclose(fid);

%%%%%%%%%%%%%%%%%  Part c  %%%%%%%%%%%%%%
% Calculate the etax function at the beginning of the first
% drift:

dsp = L*F*dsp0;
printf("For part c, we need the dispersion at the end of the"
printf(" first drift:\n")
printf("  eta_x  = %10.6f\n", dsp(1));
printf("  etap_x = %10.6f\n", dsp(2));

L = 4*lcell;
alphap = 8/L*(dsp(1)*sin(pi/4)-dsp(2)*rho*(cos(pi/4)-1)\
    +rho*(pi/4-sin(pi/4)));
printf("alpha_p = %f\n", alphap);
```

6.2.3   *Commands for* gnuplot[70]

```
#!/usr/bin/env gnuplot

set grid
set xlabel "{/Times-Italic s} [m]"
set ylabel "Betatron and dispersion functions [m]"

plot "optic.dat" us 1:2 title \
        "{/Symbol b}_{/Times-Italic x}" wi li, \
    "" us 1:3  title "{/Symbol b}_{/Times-Italic y}" wi li, \
    "" us 1:4  title "{/Symbol h}_{/Times-Italic x}" wi li, \
    "prob6-2.sloc" title "" wi li

set term post eps enh dashed lw 2 font "Times-Roman" 14 \
    size 8.5cm,5cm
set output "prob6-2.eps"
replot
unset output
```

set term wxt

6.3   Problem 6–3: Momentum compaction from a matrix

Given a full turn transfer matrix for (x, x', z, δ), i. e.,

$$\mathbf{M} = \begin{pmatrix} C & S & 0 & D \\ C' & S' & 0 & D' \\ E & F & 1 & G \\ 0 & 0 & 0 & 1 \end{pmatrix}, \tag{6.17}$$

find an expression for the momentum compaction. Assume the circumference is L, and the design momentum is p_0.

Solution:

First, we should start by noting that this result is only true in the ultra-relativistic limit $(\beta \to 1)$ More accurately, for lower energies it will lead to the phase-slip factor η_{tr} (defined at the beginning of CM: Chapter 7) rather than the momentum compaction. Problem 7–2 and § 7.4 will elaborate on the actual definition of the longitudinal canonical coordinate for the case with $\beta < 1$.

Two of the eigenvalues of $\mathbf{M}$ in Eq. (6.17) are one, and correspond to the eigenvectors

$$\begin{pmatrix} \eta \\ \eta' \\ 0 \\ 1 \end{pmatrix}, \quad \text{and} \quad \begin{pmatrix} 0 \\ 0 \\ 1 \\ 0 \end{pmatrix}, \tag{6.18}$$

where the dispersion functions are defined by Eqs. (CM: 5.87 and CM: 5.88):

$$\eta(s) = \frac{[1 - S'(s)]D(s) + S(s)D'(s)}{2(1 - \cos\mu)}, \tag{CM: 5.87}$$

$$\eta'(s) = \frac{[1 - C(s)]D'(s) + C(s)D(s)}{2(1 - \cos\mu)}, \tag{CM: 5.88}$$

with $\mu = 2\pi Q_{\mathrm{H}}$. For the closed orbit, $x_c = x_1 = x_0$ and $x'_c = x'_1 = x'_0$, so we have

$$\frac{dx_c}{d\delta} = \eta, \quad \text{and} \quad \frac{dx'_c}{d\delta} = \eta'. \tag{6.19}$$

The second eigenvector of Eq. (CM: 5.88) shows that the longitudinal coordinate of the closed orbit is independent of the momentum, as we should expect for a ring with no longitudinal focusing (i. e. for $M_{43} = 0$).

In the ultrarelativistic limit, we may write the one turn matrix as the Jacobian transformation from one turn (turn-0) to the next (turn-1):

$$
\mathbf{M} = \begin{pmatrix}
\frac{\partial x_1}{\partial x_0} & \frac{\partial x_1}{\partial x_0'} & \frac{\partial x_1}{\partial z_0} & \frac{\partial x_1}{\partial \delta_0} \\
\frac{\partial x_1'}{\partial x_0} & \frac{\partial x_1'}{\partial x_0'} & \frac{\partial x_1'}{\partial z_0} & \frac{\partial x_1'}{\partial \delta_0} \\
\frac{\partial z_1}{\partial x_0} & \frac{\partial z_1}{\partial x_0'} & \frac{\partial z_1}{\partial z_0} & \frac{\partial z_1}{\partial \delta_0} \\
\frac{\partial \delta_1}{\partial x_0} & \frac{\partial \delta_1}{\partial x_0'} & \frac{\partial \delta_1}{\partial z_0} & \frac{\partial \delta_1}{\partial \delta_0}
\end{pmatrix} ,
\tag{6.20}
$$

From the third row with a small offset from the design orbit

$$
\frac{dz_1}{d\delta} = \frac{\partial z_1}{\partial x_0} \frac{dx_0}{d\delta} + \frac{\partial z_1}{\partial x_0'} \frac{dx_0'}{d\delta} + \frac{\partial z_1}{\partial z_0} \frac{dz_0}{d\delta} + \frac{\partial z_1}{\partial \delta_0}
$$

$$
= E\eta + F\eta' + \frac{dz_0}{d\delta_0} + G,
\tag{6.21}
$$

or on rearranging and dividing by the circumference L:

$$
\alpha_p = \frac{p}{L} \frac{d\Delta L}{dp} = \frac{p}{L} \frac{d(z_1 - z_0)}{d\delta} = \frac{E\eta + F\eta' + G}{L}.
\tag{6.22}
$$

6.4   Problem 6–4: CESR luminosity

a) Calculate the luminosity in CESR, assuming $\beta_x^* = 1.1$ m, $\beta_y^* = 2$ cm, $\eta_x^* = 1.1$ m, $\sigma_p/p = 6 \times 10^{-4}$, and rms emittances of $\pi\epsilon_x = \pi \times 10^{-7}$ m and $\pi\epsilon_y = \pi \times 6 \times 10^{-8}$ m. (See Problem 1-1.)

b) What is the value if we change β_y^* to 1 cm? (You may want to do a numerical integration.)

Solution:

a) Referring to Problem 1-1, as suggested in question "a" of the problem to solved, we start with a few considerations about the values of the beta functions in the low-beta section. From Problem 1-1 we have $\sigma_z = 2.2$ cm, so $\beta_x^* = 1.1$ m $\gg \sigma_z$, and

$$\sigma_x \simeq \sigma_x^* = \sqrt{\epsilon_x \beta_x^* + \eta_x^2 \delta^2} \simeq 7.38 \times 10^{-4} \text{ m}. \tag{6.23}$$

Since the vertical β-function is comparable to the bunch length, we should use the parabolic formula

$$\beta_y(s) = \beta_y^* + \frac{s^2}{\beta_y^*}. \tag{6.24}$$

We also have $N_b = 7$ bunches and a revolution frequency (see Eq. 1.9)) of $f_0 = 391$ kHz.

Starting from Eq. (D.29) for the luminosity for a single bunch crossing with rms emittances:

$$L = \int_{-\infty}^{\infty} \frac{2N_1 N_2}{\sqrt{(2\pi)^3}} \frac{e^{-\frac{(h_x + \theta_x s)^2}{2[\epsilon_{x1}\beta_{x1}(s) + \epsilon_{x2}\beta_{x2}(s)]}}}{\sqrt{\epsilon_{x1}\beta_{x1}(s) + \epsilon_{x2}\beta_{x2}(s)}} \frac{e^{-\frac{(h_y + \theta_y s)^2}{2[\epsilon_{y1}\beta_{y1}(s) + \epsilon_{y2}\beta_{y2}(s)]}}}{\sqrt{\epsilon_{y1}\beta_{y1}(s) + \epsilon_{y2}\beta_{y2}(s)}}$$

$$\times \frac{e^{-\frac{(2s+\Delta)^2}{2(\epsilon_{z1}\beta_{z1}^* + \epsilon_{z2}\beta_{z2}^*)}}}{\sqrt{\epsilon_{z1}\beta_{z1}^* + \epsilon_{z2}\beta_{z2}^*}} \, ds,$$

$$\text{(CM: D.39)}$$

$$\rightarrow \frac{2N_1 N_2}{\sqrt{(2\pi)^3}} \int_{-\infty}^{\infty} \frac{1}{\sqrt{2\sigma_x^*}} \frac{1}{\sqrt{2\epsilon_y \beta_y(s)}} \frac{e^{-\frac{s^2}{\sigma_z^*}}}{\sqrt{2\sigma_z}} \, ds, \tag{6.25}$$

having adjusted the horizontal size σ_x for nonzero dispersion, and where we have assumed that $h_x = h_y = \theta_x = \theta_y = \Delta = 0$ for head-on collisions, as well as having the equal emittances and β functions for the two beams. Making the substitution $s = \sigma_z \xi$ and defining $r = \sigma_z / \beta_y^*$, we have

$$L = \frac{N_1 N_2}{4\pi \sigma_x^* \sigma_y^*} \frac{1}{\sqrt{\pi}} \int_{-\infty}^{\infty} \frac{e^{-\xi^2}}{\sqrt{1 + r^2 \xi^2}} \, d\xi. \tag{6.26}$$

Let us define the "hour-glass factor"

$$F(r) = \frac{1}{\sqrt{\pi}} \int_{-\infty}^{\infty} \frac{e^{-\xi^2}}{\sqrt{1 + r^2 \xi^2}} \, d\xi. \tag{6.27}$$

The formula for the total luminosity will be

$$\mathcal{L} = \frac{f_{\text{rev}} N_b N_{\pm}^2}{4\pi \sigma_x^* \sigma_y^*} F(r), \tag{6.28}$$

with $N_{\pm} = 1.28 \times 10^{11}$ from Eq. (1.10).
We have

$$\mathcal{L}_0 = \frac{3.91 \times 10^5 \, [\text{s}^{-1}] \times 7 \times (1.28 \times 10^{11})^2}{4\pi \times 7.38 \times 10^{-4} \, [\text{m}] \times \sqrt{0.02} \, [\text{m}] \times 6 \times 10^{-8} \, [\text{m}]}$$

$$= 1.40 \times 10^{31} [\text{cm}^{-2}\text{s}^{-1}], \tag{6.29}$$

and $r = 2.2/2 = 1.1$. Using Maxima[2] to evaluate the integral, we find $F(1.1)=0.8422$, so $\mathcal{L} = 1.18 \times 10^{-31} [\text{cm}^{-2}\text{s}^{-1}]$.
The two lines of Maxima code were:

```
f(x,r):=exp(-x**2)/sqrt(1+r**2*x**2)/sqrt(%pi);
quad_qags(f(x,1.1),x,-10,10);
```

Here we limited the integration to the interval $[-10, 10]$ since the integrand was essentially zero beyond these limits.

b) We need to reevaluate both σ_y and $F(2.2/1) = 0.6811$, with $r = 2.2/1$, or scaling from the previous value yields:

$$\mathcal{L} = 1.18 \times 10^{-31} [\text{cm}^{-2}\text{s}^{-1}] \times \sqrt{\frac{1 \, [\text{cm}]}{2 \, [\text{cm}]}} \times \frac{0.6811}{0.8422}$$

$$= 6.72 \times 10^{-30} \, [\text{cm}^{-2}\text{s}^{-1}]. \tag{6.30}$$

6.5 Problem 6–5: Teng-Edwards decoupling

Derive the conditions in Eqs. (CM: 6.84–6.88).

$$\cos\mu_u - \cos\mu_v = \frac{1}{2}\mathrm{tr}(\mathbf{M} - \mathbf{N})\left[1 + \frac{2\det(\mathbf{m}) + \mathrm{tr}(\mathbf{nm})}{\left[\frac{1}{2}\mathrm{tr}(\mathbf{M} - \mathbf{N})\right]^2}\right]^{\frac{1}{2}}. \quad \text{(CM: 6.84)}$$

$$\cos(2\phi) = \frac{\frac{1}{2}\mathrm{tr}(\mathbf{M} - \mathbf{N})}{\cos\mu_u - \cos\mu_v}. \quad \text{(CM: 6.85)}$$

$$\mathbf{D} = -\frac{\mathbf{m} + \tilde{\mathbf{n}}}{(\cos\mu_u - \cos\mu_v)\sin(2\phi)}. \quad \text{(CM: 6.86)}$$

$$\mathbf{A} = \mathbf{M} - \mathbf{D}^{-1}\mathbf{m}\tan\phi. \quad \text{(CM: 6.87)}$$

$$\mathbf{B} = \mathbf{N} + \mathbf{Dn}\tan\phi. \quad \text{(CM: 6.88)}$$

Solution:[25]

In 2×2-block form, recall that we have defined

$$\mathbf{T} = \begin{pmatrix} \mathbf{M} & \mathbf{n} \\ \mathbf{m} & \mathbf{N} \end{pmatrix}, \quad \text{(CM: 6.75)}$$

$$\mathbf{U} = \begin{pmatrix} \mathbf{A} & \mathbf{0} \\ \mathbf{0} & \mathbf{B} \end{pmatrix}, \quad \text{(CM: 6.76)}$$

$$\mathbf{R} = \begin{pmatrix} \mathbf{I}\cos\phi & \mathbf{D}^{-1}\sin\phi \\ -\mathbf{D}\sin\phi & \mathbf{I}\cos\phi \end{pmatrix}, \quad \text{(CM: 6.77)}$$

as well as the 2×2 matrix

$$\mathbf{D} = \begin{pmatrix} a & b \\ c & d \end{pmatrix}, \quad \text{with} \quad |\mathbf{D}| = 1. \quad \text{(CM: 6.78)}$$

Evaluating Eq. (CM: 6.80)

$$\mathbf{T} = \mathbf{RUR}^{-1} \quad \text{(CM: 6.80)}$$

$$= \begin{pmatrix} \mathbf{I}\cos\phi & \mathbf{D}^{-1}\sin\phi \\ -\mathbf{D}\sin\phi & \mathbf{I}\cos\phi \end{pmatrix} \begin{pmatrix} \mathbf{A} & \mathbf{0} \\ \mathbf{0} & \mathbf{B} \end{pmatrix} \begin{pmatrix} \mathbf{I}\cos\phi & -\mathbf{D}^{-1}\sin\phi \\ \mathbf{D}\sin\phi & \mathbf{I}\cos\phi \end{pmatrix}$$

$$= \begin{pmatrix} \mathbf{A}\cos\phi & \mathbf{D}^{-1}\mathbf{B}\sin\phi \\ -\mathbf{DA}\sin\phi & \mathbf{B}\cos\phi \end{pmatrix} \begin{pmatrix} \mathbf{I}\cos\phi & -\mathbf{D}^{-1}\sin\phi \\ \mathbf{D}\sin\phi & \mathbf{I}\cos\phi \end{pmatrix}$$

$$= \begin{pmatrix} \mathbf{A}\cos^2\phi + \mathbf{D}^{-1}\mathbf{BD}\sin^2\phi & (\mathbf{D}^{-1}\mathbf{B} - \mathbf{AD}^{-1})\sin\phi\cos\phi \\ (\mathbf{BD} - \mathbf{DA})\sin\phi\cos\phi & \mathbf{DAD}^{-1}\sin^2\phi + \mathbf{B}\cos^2\phi \end{pmatrix} \quad (6.31)$$

Comparing Eqs. (CM: 6.75 and 6.31), we have

$$\mathbf{M} = \mathbf{A}\cos^2\phi + \mathbf{D}^{-1}\mathbf{BD}\sin^2\phi, \tag{6.32}$$

$$\mathbf{N} = \mathbf{B}\cos^2\phi + \mathbf{DAD}^{-1}\sin^2\phi, \tag{6.33}$$

$$\mathbf{n} = (\mathbf{D}^{-1}\mathbf{B} - \mathbf{AD}^{-1})\sin\phi\cos\phi, \tag{6.34}$$

$$\mathbf{m} = (\mathbf{BD} - \mathbf{DA})\sin\phi\cos\phi. \tag{6.35}$$

Taking the traces of Eq. (6.32 and 6.33) yields

$$\mathrm{tr}(\mathbf{M}) = \mathrm{tr}(\mathbf{A})\cos^2\phi + \mathrm{tr}(\mathbf{B})\sin^2\phi, \tag{6.36}$$

$$\mathrm{tr}(\mathbf{N}) = \mathrm{tr}(\mathbf{B})\cos^2\phi + \mathrm{tr}(\mathbf{A})\sin^2\phi, \tag{6.37}$$

and then subtracting and dividing by two, we get

$$\frac{1}{2}\mathrm{tr}(\mathbf{M} - \mathbf{N}) = \frac{1}{2}[\mathrm{tr}(\mathbf{A}) - \mathrm{tr}(\mathbf{B})]\cos(2\phi)$$

$$= [\cos\mu_u - \cos\mu_v]\cos(2\phi), \tag{6.38}$$

which is obviously equivalent to Eq. (CM: 6.85). Combining a rearrangement of Eq. (6.34)

$$\mathbf{A} = \mathbf{D}^{-1}\mathbf{BD} - \frac{\mathbf{nD}}{\sin\phi\cos\phi}, \tag{6.39}$$

with Eq. (6.33) produces

$$\mathbf{N} = \mathbf{B}\cos^2\phi + \mathbf{D}\left(\mathbf{D}^{-1}\mathbf{BD} - \frac{\mathbf{nD}}{\sin\phi\cos\phi}\right)\mathbf{D}^{-1}\sin^2\phi$$

$$= \mathbf{B} - \mathbf{Dn}\tan\phi, \tag{6.40}$$

which is obviously equivalent to Eq. (CM: 6.88). In a likewise manner, we can solve for $\mathbf{A}$:

$$\mathbf{B} = \mathbf{DAD}^{-1} + \frac{\mathbf{mD}^{-1}}{\sin\phi\cos\phi}, \quad \text{(from 6.35)}, \tag{6.41}$$

$$\mathbf{M} = \mathbf{A}\cos^2\phi + \mathbf{A}\sin^2\phi + \mathbf{D}^{-1}\mathbf{m}\tan\phi, \quad \text{(from 6.32)}, \tag{6.42}$$

$$\mathbf{A} = \mathbf{M} - \mathbf{D}^{-1}\mathbf{m}\tan\phi. \tag{6.43}$$

Evaluating $\mathbf{m} + \tilde{\mathbf{n}}$, we have

$$\mathbf{m} + \tilde{\mathbf{n}} = -\sin\phi\cos\phi[(\mathbf{DA} - \mathbf{BD}) + \mathbf{S}(\mathbf{AD}^{-1} - \mathbf{D}^{-1}\mathbf{B})^{\mathrm{T}}\mathbf{S}^{\mathrm{T}}]$$

$$= -\frac{\sin(2\phi)}{2}[(\mathbf{DA} - \mathbf{BD}) + \mathbf{S}(\mathbf{ASD}^{\mathrm{T}}\mathbf{S}^{\mathrm{T}} - \mathbf{SD}^{\mathrm{T}}\mathbf{S}^{\mathrm{T}}\mathbf{B})^{\mathrm{T}}\mathbf{S}^{\mathrm{T}}]$$

$$= -\frac{\sin(2\phi)}{2}[(\mathbf{DA} - \mathbf{BD}) + \mathbf{S}(\mathbf{SDS}^{\mathrm{T}}\mathbf{A}^{\mathrm{T}} - \mathbf{B}^{\mathrm{T}}\mathbf{SDS}^{\mathrm{T}})\mathbf{S}^{\mathrm{T}}]$$

$$= -\frac{\sin(2\phi)}{2}[(\mathbf{DA} - \mathbf{BD}) + (\mathbf{DA}^{-1} - \mathbf{B}^{-1}\mathbf{D})]$$

$$= -\frac{\sin(2\phi)}{2}\mathbf{D}[\mathrm{tr}(\mathbf{A}) - \mathrm{tr}(\mathbf{B})], \quad \text{or}$$

$$\mathbf{D} = -\frac{\mathbf{m} + \tilde{\mathbf{n}}}{\frac{1}{2}[\mathrm{tr}(\mathbf{A}) - \mathrm{tr}(\mathbf{B})]\sin(2\phi)} \tag{6.44}$$

which produces Eq. (CM: 6.86) after we realize that $\text{tr}(\mathbf{A}) = 2\cos\mu_u$ and $\text{tr}(\mathbf{B}) = 2\cos\mu_v$. Inverting Eq. (6.44) and multiplying by $\mathbf{m}$:

$$\mathbf{D}^{-1}\mathbf{m} = -\frac{(\tilde{\mathbf{m}} + \mathbf{n})\mathbf{m}}{\frac{1}{2}[\text{tr}(\mathbf{A}) - \text{tr}(\mathbf{B})]\sin(2\phi)}$$
$$= -\frac{\mathbf{I}|\mathbf{m}| + \mathbf{nm}}{\frac{1}{2}[\text{tr}(\mathbf{A}) - \text{tr}(\mathbf{B})]\sin(2\phi)}, \tag{6.45}$$

but from Eq. (6.35) we also have

$$\mathbf{D}^{-1}\mathbf{m} = (\mathbf{D}^{-1}\mathbf{B}\mathbf{D} - \mathbf{A})\sin\phi\cos\phi \tag{6.46}$$

Equating the traces of the last two equations yields

$$\text{tr}(\mathbf{A} - \mathbf{B})\sin\phi\cos\phi = \frac{\text{tr}(\mathbf{nm}) + 2|\mathbf{m}|}{\frac{1}{2}\text{tr}(\mathbf{A} - \mathbf{B})\sin(2\phi)}, \tag{6.47}$$

$$\left[\frac{1}{2}\text{tr}(\mathbf{A} - \mathbf{B})\right]^2 = \frac{\text{tr}(\mathbf{nm}) + 2|\mathbf{m}|}{1 - \cos^2(2\phi)} = \frac{\text{tr}(\mathbf{nm}) + 2|\mathbf{m}|}{1 - \left(\frac{\frac{1}{2}\text{tr}(\mathbf{M}-\mathbf{N})}{\frac{1}{2}\text{tr}(\mathbf{A}-\mathbf{B})}\right)^2}, \tag{6.48}$$

$$\left[\frac{1}{2}\text{tr}(\mathbf{A} - \mathbf{B})\right]^2 = \left[\frac{1}{2}\text{tr}(\mathbf{M} - \mathbf{N})\right]^2 + \text{tr}(\mathbf{nm}) + 2|\mathbf{m}|. \tag{6.49}$$

Taking the square root gives the desired Eq. (CM: 6.84).

6.6 Problem 6–6: Determinant of a symplectic matrix

Show that any real symplectic $2n \times 2n$ matrix has a determinant of $+1$.

Note: This result means that there are no "improper" matrices in the symplectic group as there are for improper rotations in the rotation group $O(n)$.

(Here we have revised the "Note" to the problem from the somewhat mangled original.)

Solution:

It follows from Brioschi's theorem[11, 54]: For any square real matrix $\mathbf{M}$ and an antisymmetric matrix $\mathbf{A}$, then

$$\text{Pf}(\mathbf{MAM^T}) = \det(\mathbf{M})\,\text{Pf}(\mathbf{A}). \qquad (6.50)$$

Taking $\mathbf{A} = \mathbf{S}$, it immediately follows that $\det(\mathbf{M}) = +1$, since $\mathbf{MSM^T} = \mathbf{S}$ and $\text{Pf}(\mathbf{S}) \neq 0$. (In fact $\text{Pf}(\mathbf{S}) = +1$.)

See Dragt[23] for a considerably longer proof without pfaffians.

Comments about the pfaffian:

The function $\text{Pf}(\mathbf{A})$ called the pfaffian[39, 8] of an $n \times n$ antisymmetric (i. e., *skew symmetric* to mathematicians) matrix $\mathbf{A}$. the pfaffian $\text{Pf}(\mathbf{A})$ is essentially the square root[13, 39] of the determinant of $\mathbf{A}$.

For odd n, we have $|\mathbf{A^T}| = |-\mathbf{A}| = (-1)^n|\mathbf{A}| = -|\mathbf{A}|$, so we find that the determinant of an antisymmetric matrix with n odd must be zero.

For an even order $2n \times 2n$ antisymmetric matrix $\mathbf{A}$, the pfaffian is defined by

$$\text{Pf}(\mathbf{A}) = \frac{1}{2^n n!} \sum_{\sigma \in S_{2n}} \text{sgn}(\sigma) \prod_{i=1}^{n} a_{\sigma(2i-1),\sigma(2i)}, \qquad (6.51)$$

where S_{2n} is the symmetric group of permutations of the $2n$ subscripts. The function sgn is the sign of the permutation.

Rather than writing a long chapter on the theory of pfaffians, we refer the reader to other sources[39, 62, 8], as well as to searching Wikipedia for "pfaffian".

6.7 Problem 6–7: Correction of natural chromaticity

Calculate the natural chromaticities of the lattice designed in Problem 6.2, for both horizontal and vertical motion. Use two families of thin sextupole magnets to cancel the natural chromaticity of the lattice with the residual chromaticity of the sextupoles. Where and how large are the sextupoles?

Solution:

Recall Eqs.(CM: 6.126 and CM: 6.127b) for the natural chromaticities:

$$\xi_{xN} = -\frac{1}{4\pi Q_{\mathrm{H}}} \oint \beta_x(s)k_0(s)\,ds, \qquad \text{(CM: 6.126)}$$

$$\xi_{yN} = \frac{1}{4\pi Q_{\mathrm{V}}} \oint \beta_y(s)k_0(s)\,ds. \qquad \text{(CM: 6.127b)}$$

For the thin lens quadrupoles these transform to

$$\xi_{xN} = -\frac{1}{4\pi Q_{\mathrm{H}}} \sum_{j\in\{\text{quads}\}} \beta_{xj}(k_0 l)_j, \qquad (6.52)$$

$$\xi_{yN} = \frac{1}{4\pi Q_{\mathrm{V}}} \sum_{j\in\{\text{quads}\}} \beta_{yj}(k_0 l)_j. \qquad (6.53)$$

The values of the β-functions at the quadrupoles in the periodic cell (as read from the output file "optic.dat" of the program listed in § 6.2.2) are

$$\beta_{x,1} = 15.494\text{m}, \qquad \beta_{y,1} = 20.857\text{m}, \qquad (6.54)$$

$$\beta_{x,2} = 7.341\text{m}, \qquad \beta_{y,2} = 37.245\text{m}, \qquad (6.55)$$

and the respective $k_0 l$ values are 0.0507 and -0.0609 m^{-1}. Calculating the chromaticities and remembering that there are 4 periodic cells:

$$\xi_{xN} = \frac{4}{4\pi \times 1.1576}[15.494 \times 0.507 + 7.341 \times (-0.0609)]$$
$$= 0.0932, \qquad (6.56)$$

$$\xi_{yN} = \frac{4}{4\pi \times 0.4601}[20.857 \times 0.507 + 37.245 \times (-0.0609)]$$
$$= -0.8368, \qquad (6.57)$$

For the best results, we would like to place horizontal sextupoles where β_x is large, i. e. near the horizontally focusing quadrupoles. Likewise, the vertical family should be placed near the vertically focusing quadrupoles where β_y is large. So if we locate the sextupoles right next to the quadrupoles, using the same s coordinates for simplicity, we have residual chromaticities from Eqs.(CM: 6.130 and CM: 6.133),

$$\xi_{xR} = \frac{1}{4\pi Q_{\mathrm{H}}} \frac{q}{p_0} \oint \beta_x(s) 2b_2(s) \eta_x(s) \, ds, \qquad \text{(CM: 6.130)}$$

$$\xi_{yR} = -\frac{1}{4\pi Q_{\mathrm{V}}} \frac{q}{p_0} \oint \beta_y(s) 2b_2(s) \eta_x(s) \, ds, \qquad \text{(CM: 6.133)}$$

where we have added the missing subscripts. These two equations become sums for thin sextupoles:

$$\xi_{xR} = \frac{N_{\mathrm{cell}}}{2\pi Q_{\mathrm{H}}} [\beta_{x1}\eta_{x1}A + \beta_{x2}\eta_{x2}B], \qquad (6.58)$$

$$\xi_{yR} = \frac{N_{\mathrm{cell}}}{2\pi Q_{\mathrm{V}}} [\beta_{y1}\eta_{x1}A + \beta_{y2}\eta_{x2}B], \qquad (6.59)$$

where the sextupole strengths A and B are respectively

$$A = \frac{q}{p_0} b_2(s_1) l_{\mathrm{sex}}, \qquad (6.60)$$

$$B = \frac{q}{p_0} b_2(s_2) l_{\mathrm{sex}}, \qquad (6.61)$$

for thin sextupoles of length l_{sex}, and where $s_1 = 0$ and $s_2 = 9.854$ m. So we must solve the linear system

$$\begin{pmatrix} \frac{N_{\mathrm{cell}}}{2\pi Q_{\mathrm{H}}}\beta_{x1}\eta_{x1} & \frac{N_{\mathrm{cell}}}{2\pi Q_{\mathrm{H}}}\beta_{x2}\eta_{x2} \\ \frac{N_{\mathrm{cell}}}{2\pi Q_{\mathrm{V}}}\beta_{y1}\eta_{x1} & \frac{N_{\mathrm{cell}}}{2\pi Q_{\mathrm{V}}}\beta_{y2}\eta_{x2} \end{pmatrix} \begin{pmatrix} A \\ B \end{pmatrix} = -\begin{pmatrix} 0.0932 \\ -0.8368 \end{pmatrix}. \qquad (6.62)$$

Evaluating the 2×2 matrix and solving, we find

$$\begin{pmatrix} A \\ B \end{pmatrix} = -\begin{pmatrix} 94.190 & 33.553 \\ 319.006 & 428.301 \end{pmatrix}^{-1} \begin{pmatrix} 0.0932 \\ -0.8368 \end{pmatrix}$$

$$= \begin{pmatrix} -0.00229 \ \mathrm{m}^{-2} \\ 0.00366 \ \mathrm{m}^{-2} \end{pmatrix}. \qquad (6.63)$$

6.8 Problem 6-8: Chromaticity for FODO cell

Consider a ring made of N identical FODO cells with equally spaced quadrupoles. Assume that the two quadrupoles are both of length l_q, but their strengths may differ. Calculate the natural chromaticities for this machine, and show that for short quadrupoles,

$$\xi_N \simeq -\frac{2\tan\frac{\mu}{2}}{\mu}, \qquad (6.64)$$

where μ is the betatron phase advance per cell.

Solution:

Using the thin lens approximation for the ODOF cell we get the matrix

$$
\begin{aligned}
\mathbf{M}_{\text{cell}} &= \begin{pmatrix} 1 & 0 \\ -\frac{1}{f_F} & 1 \end{pmatrix} \begin{pmatrix} 1 & l \\ 0 & 1 \end{pmatrix} \begin{pmatrix} 1 & 0 \\ \frac{1}{f_D} & 1 \end{pmatrix} \begin{pmatrix} 1 & l \\ 0 & 1 \end{pmatrix} \\
&= \begin{pmatrix} 1 & l \\ -\frac{1}{f_F} & 1-\frac{l}{f_F} \end{pmatrix} \begin{pmatrix} 1 & l \\ \frac{1}{f_D} & 1+\frac{l}{f_D} \end{pmatrix} \\
&= \begin{pmatrix} 1+\frac{l}{f_D} & 2l+\frac{l^2}{f_D} \\ -\frac{1}{f_F}+\frac{1}{f_D}-\frac{l}{f_F f_D} & -\frac{l}{f_F}+\left[1-\frac{l}{f_F}\right]\left[1+\frac{l}{f_D}\right] \end{pmatrix} \\
&= \begin{pmatrix} 1+\frac{l}{f_D} & 2l+\frac{l^2}{f_D} \\ \frac{f_F-f_D-l}{f_F f_D} & 1-\frac{2l}{f_F}+\frac{l}{f_D}-\frac{l^2}{f_F f_D} \end{pmatrix},
\end{aligned} \qquad (6.65)
$$

with

$$\cos\mu = \frac{1}{2}\text{tr}(\mathbf{M}_{\text{cell}}) = 1+\frac{l}{f_D}-\frac{l}{f_F}-\frac{l^2}{2f_D f_F} = 1-2\sin^2\frac{\mu}{2}, \qquad (6.66)$$

or

$$2\sin^2\frac{\mu}{2} = \frac{l}{f_F}-\frac{l}{f_D}+\frac{l^2}{2f_D f_F}. \qquad (6.67)$$

The β_{max} will occur at the focusing quadrupole. Since Eq. (6.65) is for a periodic cell starting at the focusing quad, the $\mathbf{M}_{12}$ component gives us

$$\beta_{\text{max}}\sin\mu = 2l+\frac{l^2}{f_D}, \qquad (6.68)$$

or upon rearranging

$$\beta_{\max} = \frac{2l + \frac{l^2}{f_D}}{\sin\mu}. \tag{6.69}$$

Interchanging f_F with f_D for a OFOD cell gives

$$\beta_{\min} = \frac{2l - \frac{l^2}{f_F}}{\sin\mu}. \tag{6.70}$$

The chromaticity is

$$\begin{aligned}
\xi_N &= -\frac{1}{2N_{\text{cell}}\mu} \oint \beta(s)k(s)ds \\
&= -\frac{1}{2N_{\text{cell}}\mu} \times N_{\text{cell}} \int_{\text{cell}} \beta(s)k(s)ds \\
&= -\frac{1}{2\mu}\left[\beta_{\max}\left(\frac{1}{f_F}\right) + \beta_{\min}\left(-\frac{1}{f_D}\right)\right] \\
&= -\frac{1}{2\mu\sin\mu}\left[\left(2l + \frac{l^2}{f_D}\right)\frac{1}{f_F} - \left(2l - \frac{l^2}{f_F}\right)\frac{1}{f_D}\right] \\
&= -\frac{1}{\mu\sin\mu}\left[\frac{l}{f_F} - \frac{l}{f_D} + \frac{l^2}{f_F f_D}\right] \\
&\simeq -\frac{1}{\mu\sin\mu}\frac{l^2}{f_F f_D}.
\end{aligned} \tag{6.71}$$

If $f_F \simeq f_D$, we have

$$\begin{aligned}
\xi_N &\simeq -\frac{1}{2\mu\sin\frac{\mu}{2}\cos\frac{\mu}{2}}4\sin^2\frac{\mu}{2} \\
&= -\frac{2\tan\frac{\mu}{2}}{\mu},
\end{aligned} \tag{6.72}$$

with help from Eq. (6.67).

Problems of Chapter 7: Synchrotron Oscillations

7.1 Problem 7–1: Longitudinal scaling relations

Show that for a ring with $\eta_{\text{tr}} = \text{constant}$, the following scaling relations hopefully apply:

$$\varphi_m(U) = \left(\frac{U_i}{U}\right)^{\frac{1}{4}} \varphi_m(U_i), \tag{7.1}$$

$$W_m(U) = \left(\frac{U}{U_i}\right)^{\frac{1}{4}} W_m(U_i), \quad \text{and} \tag{7.2}$$

$$\left(\frac{\Delta p}{p}\right) = \left[\frac{(\gamma_i^2 - 1)^2 \gamma}{(\gamma^2 - 1)^2 \gamma_i}\right]^{\frac{1}{4}} \left(\frac{\Delta p}{p}\right)_i, \tag{7.3}$$

where the subscript i indicates the initial conditions.

Solution:

Solving

$$W_m = \frac{\Omega_s \beta^2 U_s}{\omega_{\text{rf}}{}^2 \eta_{\text{tr}}} \varphi_m, \tag{CM: 7.90}$$

for φ_m, and substituting into the longitudinal invariant

$$I_L = \cdots = \pi \varphi_m W_m, \tag{CM: 7.91}$$

gives rise to

$$I_L = \frac{\pi \omega_{\text{rf}}^2 \eta_{\text{tr}}}{\Omega_s \beta^2 U_s} W_m^2. \tag{7.4}$$

Squaring this expression, and bearing in mind that square of the synchrotron oscillation frequency is

$$\Omega_s^2 = \frac{qV h \eta_{\mathrm{tr}} \omega_s^2 \cos \phi_s}{2\pi \beta^2 U_s}, \qquad \text{(CM: 7.84)}$$

we obtain

$$\begin{aligned}
W_m^4 &= \frac{qV \cos \phi_s \, \eta_{\mathrm{tr}} h \omega_s^2}{2\pi \beta^2 U_s} \frac{\beta^4 U_s^2}{\pi^2 \omega_{\mathrm{rf}}{}^4 \eta_{\mathrm{tr}}{}^2} I_L^2 \\
&= \frac{qV \cos \phi_s \, \beta^2 U_s}{2\pi^3 h \omega_{\mathrm{rf}}{}^2 \eta_{\mathrm{tr}}} I_L^2,
\end{aligned} \qquad (7.5)$$

which, multiplied by $\pi^4 \varphi_m^4$, yields the fourth power of the invariant

$$I_L^4 = (\pi \varphi_m W_m)^4 = \frac{\pi qV \cos \phi_s \, \beta^2 U_s}{2 h \omega_{\mathrm{rf}}{}^2 \eta_{\mathrm{tr}}} \varphi_m^4 I_L^2, \qquad (7.6)$$

or

$$\varphi_m^2 = \frac{\omega_{\mathrm{rf}}}{\beta} \sqrt{\frac{2 h \eta_{\mathrm{tr}}}{\pi qV \cos \phi_s}} \frac{I_L^2}{\sqrt{U_s}}. \qquad (7.7)$$

On the other hand

$$\frac{\omega_{\mathrm{rf}}}{\beta} = \frac{h \omega_s}{\beta} = \frac{2\pi h c}{\beta c T_s} = \frac{2\pi h c}{L}, \qquad (7.8)$$

where L is the ring circumference. Hence we have:

$$\varphi_m^2 = \frac{2\pi h c}{L} \sqrt{\frac{2 h \eta_{\mathrm{tr}}}{\pi qV \cos \phi_s}} \frac{I_L^2}{\sqrt{U_s}}, \qquad (7.9)$$

which gives rise to the following relations, energy changes adiabatically, and that U is far enough above transition that η_{tr} is essentially constant:

$$\varphi_m(U) U^{\frac{1}{4}} = \varphi_m(U_i) U_i^{\frac{1}{4}} = \text{constant}, \qquad (7.10)$$

and

$$\varphi_m(U) = \left(\frac{U_i}{U}\right)^{\frac{1}{4}} \varphi_m(U_i), \qquad \text{or} \qquad \varphi_m(\gamma) = \left(\frac{\gamma_i}{\gamma}\right)^{\frac{1}{4}} \varphi_m(\gamma_i). \qquad (7.11)$$

Using again the invariant I_L we obtain:

$$\pi \varphi_m(U) W_m(U) = \pi \varphi_m(U_i) U_i^{\frac{1}{4}} W_m(U) = \pi \varphi_m(U_i) W_m(U_i), \qquad (7.12)$$

and

$$W_m(U) = \left(\frac{U}{U_i}\right)^{\frac{1}{4}} W_m(U_i), \qquad \text{or} \qquad W_m(\gamma) = \left(\frac{\gamma}{\gamma_i}\right)^{\frac{1}{4}} W_m(\gamma_i). \qquad (7.13)$$

In order to answer to find the third relation, we need to write

$$W = -\frac{\delta U}{\omega_{\text{rf}}}, \qquad \text{(CM: 7.58)}$$

in terms of $\delta = \Delta p/p$. Recalling

$$\frac{dU}{U} = \frac{p^2}{U^2}\frac{dp}{p} = \beta^2\frac{dp}{p}, \qquad \text{(CM: 1.17)}$$

so

$$W = -\frac{\delta U}{\omega_{\text{rf}}} = -\frac{\beta^2\gamma mc}{\omega_{\text{rf}}}\frac{\Delta p}{p}. \qquad (7.14)$$

or, on substitution of Eq. (7.8),

$$W(\gamma) = -\frac{Lmc}{2\pi h}\gamma\beta\left(\frac{\Delta p}{p}\right), \qquad (7.15)$$

This may be transformed into

$$\left(\frac{\Delta p}{p}\right) = -\frac{2\pi h}{Lmc}(\gamma^2 - 1)^{-\frac{1}{2}}W(\gamma), \qquad (7.16)$$

$$\left(\frac{\Delta p}{p}\right)_i = -\frac{2\pi h}{Lmc}(\gamma_i^2 - 1)^{-\frac{1}{2}}W(\gamma_i), \qquad (7.17)$$

whose ratio yields

$$\left(\frac{\Delta p}{p}\right)\bigg/\left(\frac{\Delta p}{p}\right)_i = \frac{(\gamma_i^2 - 1)^{\frac{1}{2}}}{(\gamma^2 - 1)^{\frac{1}{2}}}\frac{W(\gamma)}{W(\gamma_i)} = \frac{(\gamma_i^2 - 1)^{\frac{1}{2}}}{(\gamma^2 - 1)^{\frac{1}{2}}}\left(\frac{\gamma}{\gamma_i}\right)^{\frac{1}{4}}, \qquad (7.18)$$

or

$$\left(\frac{\Delta p}{p}\right) = \left[\frac{(\gamma_i^2 - 1)^2}{(\gamma^2 - 1)^2}\frac{\gamma}{\gamma_i}\right]^{\frac{1}{4}}\left(\frac{\Delta p}{p}\right)_i \simeq \left(\frac{\gamma_i}{\gamma}\right)^{\frac{3}{4}}\left(\frac{\Delta p}{p}\right)_i \qquad \text{(UR),} \quad (7.19)$$

having taken into account Eq. (7.13).

Comment:

The hypothesis of having a constant η_{tr} deserves a few comments. In fact, recalling the definition (CM: 7.1)

$$\eta_{\text{tr}} = \frac{1}{\gamma^2} - \alpha_{\text{p}}, \qquad \text{(CM: 7.1)}$$

we can state that this hypothesis is surely correct for ultrarelativistic particles when $\gamma \gg 1$ and consequently $\eta_{\text{tr}} \simeq -\alpha_{\text{p}}$. For intermediate energies we must consider the tangible dependence on $U = \gamma mc^2$. Since in all our expressions U and η_{tr} appear always as a quotient, we can easily modify Eqs. (7.11)and (7.13) respectively into

$$\varphi_m(\gamma) = \left[\frac{1 - \alpha_p \gamma^2}{1 - \alpha_p \gamma_i^2} \left(\frac{\gamma_i}{\gamma}\right)^3\right]^{\frac{1}{4}} \varphi_m(\gamma_i), \qquad (7.20)$$

$$W_m(\gamma) = \left[\frac{1 - \alpha_p \gamma_i^2}{1 - \alpha_p \gamma^2} \left(\frac{\gamma}{\gamma_i}\right)^3\right]^{\frac{1}{4}} W_m(\gamma_i), \qquad (7.21)$$

having left aside Eq. (7.19) since it is rather complicated and time consuming demonstration that does not deserve attention. However, in the weak focusing synchrotrons example, the constant η_{tr} hypothesis is quite reasonable. In fact, let us consider e.g. the Frascati electron synchrotron, dismantled many years ago, and let us recall that by U_i=2.5 MeV+0.5 MeV ($\gamma_i = 6$), $U_{\text{Max}} = 1$ GeV (γ_{Max}) = 2000, $\rho = 3.60$ m, $l_0 = 1.20$ m, $n = 0.61$. Inserting these parameters into the expression of the momentum compaction factor

$$\alpha_p = \frac{dL/L}{dp/p} = \left(1 + \frac{2l_0}{\pi\rho}\right)^{-1} (1 - n)^{-1} \simeq \frac{1}{Q_H^2}, \qquad \text{(CM: 2.79)}$$

which is typical of any four-quadrant weak-focusing synchrotron, we obtain:

$$\alpha_p = \frac{1}{1 + \frac{2l_0}{\pi\rho}} \frac{1}{1 - n} = 2.115, \qquad (7.22)$$

and

$$(\eta_{tr})^{(i)} = 0.028 - 2.115 = 2.087, \qquad (7.23)$$

exhibiting, even at the injection, a discrepancy $\Delta\eta/\eta$ of the order of 1%. Besides, this discrepancy will decrease with the growing up of the machine energy. However, in electron rings, everything dealing with the size of any emittance is strongly dependent on the synchrotron radiation.

7.2   Problem 7–2: Dispersion function with rf

Consider a ring with a thin rf cavity whose linear transfer matrix just after the cavity is given by

$$M = \begin{pmatrix} C & S & 0 & D \\ C' & S' & 0 & D' \\ E & F & 1 & G \\ 0 & 0 & 0 & 1 \end{pmatrix} \begin{pmatrix} 1 & 0 & 0 & 0 \\ 0 & 1 & 0 & 0 \\ 0 & 0 & 1 & 0 \\ 0 & 0 & Q & 1 \end{pmatrix}. \tag{7.24}$$

Show that the dispersion functions are still given by

$$\eta = \frac{(1 - S')D + SD'}{2(1 - \cos\mu)}, \quad \text{and} \quad \eta' = \frac{(1 - C)D' + C'D}{2(1 - \cos\mu)}. \tag{7.25}$$

Hint: The eigenvector equation of Eq. 5.86 must be modified to allow for momentum compaction:

$$M \begin{pmatrix} \eta \\ \eta' \\ 0 \\ 1 \end{pmatrix} \delta = \begin{pmatrix} \eta \\ \eta' \\ 0 \\ 1 \end{pmatrix} \delta + \begin{pmatrix} 0 \\ 0 \\ \Delta L \\ 0 \end{pmatrix}. \tag{7.26}$$

Solution:

Multiplying out the matrices in Eq. (7.24) results in

$$M = \begin{pmatrix} C & S & 0 & D \\ C' & S' & 0 & D' \\ E & F & 1 & G \\ 0 & 0 & 0 & 1 \end{pmatrix} \begin{pmatrix} 1 & 0 & 0 & 1 \\ 0 & 1 & 0 & 0 \\ 0 & 0 & 1 & 0 \\ 0 & 0 & Q & 1 \end{pmatrix} = \begin{pmatrix} C & S & DQ & D \\ C' & S' & D'Q & D' \\ E & F & 1+GQ & G \\ 0 & 0 & Q & 1 \end{pmatrix}. \tag{7.27}$$

In CM: §5.5 we solved for the dispersion functions by finding the eigenvector of the matrix

$$\begin{pmatrix} C & S & D \\ C' & S' & D' \\ 0 & 0 & 1 \end{pmatrix} \tag{7.28}$$

corresponding to the eigenvalue $\lambda = 1$; however we cannot so simply write down such a matrix, since the elements of the third column of M which were previously zeros are now nonzero. In general, if the rf cavity is located in a dispersive straight section (with either D or D' unequal to zero), then it may be that none of the eigenvalues of M is 1. In fact, for stable motion in both the horizontal and longitudinal planes, both the betatron and the synchrotron tunes differ from integers, thus ensuring that none of the eigenvalues is $+1$.

Substituting Eq. (7.27) into Eq. (7.26) and rearranging gives

$$\begin{pmatrix} C-1 & S & DQ & D \\ C' & S'-1 & D'Q & D \\ E & F & GQ & G \\ 0 & 0 & Q & 0 \end{pmatrix} \begin{pmatrix} \eta \\ \eta' \\ 0 \\ 1 \end{pmatrix} = \begin{pmatrix} 0 \\ 0 \\ \frac{\Delta L}{\delta} \\ 0 \end{pmatrix}. \tag{7.29}$$

The top two rows are independent of Q:

$$\begin{pmatrix} C-1 & S \\ C' & S'-1 \end{pmatrix} \begin{pmatrix} \eta \\ \eta' \end{pmatrix} + \begin{pmatrix} D \\ D' \end{pmatrix} = \begin{pmatrix} 0 \\ 0 \end{pmatrix}, \tag{7.30}$$

and can be solved as in CM: §5.5

$$\begin{pmatrix} \eta \\ \eta' \end{pmatrix} = -\begin{pmatrix} C-1 & S \\ C' & S'-1 \end{pmatrix}^{-1} \begin{pmatrix} D \\ D' \end{pmatrix}$$

$$= -\frac{1}{(C-1)(S'-1)-SC'} \begin{pmatrix} S'-1 & -S \\ -C' & C-1 \end{pmatrix} \begin{pmatrix} D \\ D' \end{pmatrix}$$

$$= \frac{1}{2(1-\cos\mu)} \begin{pmatrix} (1-S')D + SD' \\ C'D + (1-C)D' \end{pmatrix}, \tag{7.31}$$

which coincides with Eq. (7.25).

The third row of Eq. (7.29) yields the required shift in the longitudinal coordinate:

$$\Delta L = (E\eta + F\eta' + G)\delta. \tag{7.32}$$

Comment:

One should note that this ΔL is not really the change in circumference of the ring except in the ultrarelativistic limit ($\beta \to 1$). The canonical longitudinal coordinate is actually $z = -\beta_0 c\Delta t$, so we may write the phase-slip factor as

$$\eta_{\text{tr}} = -\frac{d\tau}{\tau} \bigg/ \frac{dp}{p} = \frac{\Delta L}{\beta_0 c\tau} \bigg/ \frac{dp}{p}$$

$$= \frac{E\eta + F\eta' + G}{L}, \tag{7.33}$$

where τ is the revolution period, and $L = \beta_0 c\tau$ is the actual circumference of the design orbit as discussed in § 7.4.

7.3 Problem 7–3: RHIC longitudinal parameters with Au

a) Calculate the synchrotron tune for RHIC for fully stripped $^{197}\text{Au}^{79+}$ (gold ions)

$$\gamma_{\text{inj}} = 10.4 \tag{7.34}$$

$$\gamma_{\text{tr}} = 22.8 \qquad \text{[correction]} \tag{7.35}$$

$$L = 3834 \text{ m} \tag{7.36}$$

$$h = 360 \tag{7.37}$$

$$\phi_s = 0° \tag{7.38}$$

$$mc^2 = 197 \times 0.93113 \text{ GeV} \tag{7.39}$$

$$Z = 79 \quad \text{(protons)} \tag{7.40}$$

$$A = 197 \quad \text{(neutrons + protons)} \qquad \text{[correction]} \tag{7.41}$$

$$V_{\text{rf}} = 300 \text{ kV}. \tag{7.42}$$

b) What is the synchrotron frequency?

c) For a synchronous phase of $\phi_s = 5.5°$, how much energy does the synchronous particle gain per turn?

d) How long would it take to accelerate to $\gamma = 107.4$ (100 GeV/nucleon)? Assume that the phase jump at transition has been performed correctly (i.e., ignore it).

e) Plot the synchrotron frequency as a function of energy.

Solution:

a) Recalling the formula for the synchrotron frequency

$$\Omega_s = \omega_s \sqrt{\frac{h\eta_{\text{tr}} \cos \phi_s}{2\pi\beta^2\gamma} \frac{qV}{mc^2}} \tag{CM:7.32}$$

we may write the tune as $Q_s = \Omega_s/\omega_s$, so we need to evaluate the square root. Since we need to answer several questions, let us first evaluate a number of the extra quantities separately:

$$\beta_{\text{inj}} = \sqrt{1 - \gamma_{\text{inj}}^{-2}} = 0.99573, \tag{7.43}$$

$$\omega_s = 2\pi\frac{\beta c}{L} = 2\pi \times 0.99537 \times \frac{2.99792 \times 10^8 \text{ [m/s]}}{3834 \text{ [m]}}$$

$$= 4.8903 \times 10^5 \text{ [radian/m]}, \tag{7.44}$$

$$f_{s,\text{inj}} = \frac{\omega_s}{2\pi} = 77.831 \text{ [kHz]}, \qquad (7.45)$$

$$\eta_{\text{tr}} = \gamma_{\text{inj}}^{-2} - \gamma_{\text{tr}}^{-1} = 0.0073219, \qquad (7.46)$$

$$mc^2 = 183.43 \text{ [GeV]}. \qquad (7.47)$$

$$Q_s = \sqrt{\frac{360 \times 0.0073219 \times 1}{2\pi \times (0.99573)^2 \times 10.4} \times \frac{79 \times 3 \times 10^5 \text{ [eV]}}{197 \times 0.93113 \text{ [GeV]}}}$$

$$= 0.0022936. \qquad (7.48)$$

b)

$$\Omega_s = Q_s \omega_s = 0.0022936 \times 4.8903 \times 10^5 = 1121.6 \text{ [radians/s]}. \quad (7.49)$$

Dividing by 2π gives

$$f_{\text{sync}} = 77831 \text{ kHz}. \qquad (7.50)$$

c)

$$\Delta U = ZeV_{\text{rf}} \sin\phi_s = 2.2715 \text{ [MeV/turn]}. \qquad (7.51)$$

d) The number of turns (assuming a constant ramp rate of the energy) would be

$$N = \frac{(\gamma_{\text{top}} - \gamma_{\text{inj}})mc^2}{\Delta U} = \frac{(107.4 - 10.4) \times 183.43 \text{ [GeV]}}{2.2715 \text{ [MeV]}}$$

$$= 7.8330 \times 10^6 \text{turns}. \qquad (7.52)$$

Calculating the revolution frequency at the top energy:

$$\beta_{\text{top}} = \sqrt{1 - (107.4)^{-2}} = 0.99996, \qquad (7.53)$$

$$f_{s,\text{top}} = \frac{\beta_{\text{top}}}{\beta_{\text{inj}}} f_{s,\text{inj}} = 78.190 \text{ kHz}, \qquad (7.54)$$

which is only slightly larger than the frequency at injection. The acceleration time is

$$t_{\text{acc}} \simeq \frac{N}{f_s} = \frac{7.833 \times 10^6}{78[\text{kHz}]} \simeq 100 \text{ s}. \qquad (7.55)$$

e) Using your favorite plotting package (MacKay's is gnuplot[70].) you should get something like Fig 7.1.

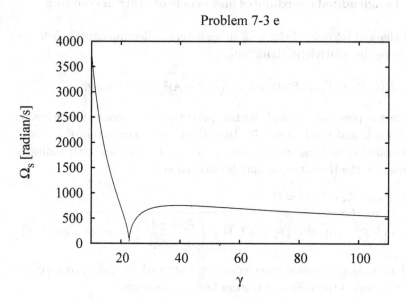

Fig. 7.1   Plot of the RHIC synchrotron frequency as a function of γ.

7.4 Longitudinal coordinates and synchrobetatron coupling

For a charged particle of charge q in an external electromagnetic field, we may write the relativistic Hamiltonian as

$$\mathcal{H}(x, P_x, y, P_y, z, P_z; t) = U = \sqrt{(\vec{P} - \vec{A})^2 + m^2 c^4} + q\Phi, \qquad (7.56)$$

with vector potential $\vec{A}$, and electric potential Φ, canonical momentum $\vec{P} = \vec{p} + q\vec{A}$, and total energy U. Here the kinetic momentum $\vec{p} = \gamma\vec{\beta}mc$. In the usual cylindrical coordinates of accelerator physics with radius of curvature ρ, the Hamiltonian may be written as

$$\mathcal{H}(x, P_x, y, P_y, s, P_s; t) = U$$

$$= c\sqrt{(P_x - qA_x)^2 + (P_y - qA_y)^2 + \left(\frac{P_s - qA_s}{1 + x/\rho}\right)^2 + m^2 c^2} + q\Phi. \quad (7.57)$$

Recalling that a canonical transformation from variables $(\vec{q}, \vec{p})$ to variables $(\vec{Q}, \vec{P})$ preserves the Poincaré-Cartan integral invariant

$$\vec{p} \cdot d\vec{q} - H\, dt = \vec{P} \cdot d\vec{Q} - K\, dt, \qquad (7.58)$$

we can interchange one canonical pair (q_j, p_j) with the time-energy pair $(t, -H)$ by writing the invariant as

$$\left(\sum_{i \neq j} p_i dq_i + (-H)dt\right) - (-p_j)dq_j. \qquad (7.59)$$

This transformation gives the new Hamiltonian

$$H(x, P_x, y, P_y, t, -U; s) = -P_s$$

$$= -qA_s - \left(1 + \frac{x}{\rho}\right)$$

$$\times \sqrt{\left(\frac{U - q\Phi}{c}\right)^2 - (mc)^2 - (P_x - qA_x)^2 - (P_y - qA_y)^2}. \quad (7.60)$$

If there are no electrostatic fields then we may write $\Phi = 0$; the fields in rf cavities may be obtained from the time derivative of $\vec{A}$. Ignoring solenoids for now, with only transverse magnetic guide fields and the longitudinal electric fields of the cavities may be described by A_s, so we may take

$$A_x = 0, \quad A_y = 0, \quad \text{and} \quad \Phi = 0. \qquad (7.61)$$

To model dipoles, quadrupoles and cavities a vector potential of the form

$$qA_s = q \left(1 + \frac{x}{\rho}\right) (\vec{A} \cdot \hat{s})$$

$$= -\frac{p_{sy}}{\rho} x - \frac{p_{sy} K}{2} (x^2 - y^2) + \cdots$$

$$+ \frac{qV}{\omega_{rf}} \sum_{j=-\infty}^{\infty} \delta(s - jL) \cos(\omega_{rf} t + \phi_0) \tag{7.62}$$

is sufficient. Here the circumference is L, and the magnetic guide field parameter is

$$K = \frac{1}{\rho^2} + \frac{q}{p_{sy}} \left(\frac{\partial B_y}{\partial x}\right)_0, \tag{7.63}$$

and p_{sy} is the momentum of the synchronous design particle. The effective rf phase as the synchronous particle passes the cavities is ϕ_0, to give a net energy gain per turn of $[qV \cos(\phi_0)]$. For simplicity in Eq. (7.62) the effect of all rf cavities has been lumped at the location $s = 0$ in the ring.

The time coordinate may be broken up into the time for the synchronous particle to arrive at the location s plus a deviation Δt for the particular particle's arrival time:

$$t = t_{sy}(s) + \Delta t(s) = \frac{2\pi h}{\omega_{rf} L} s + \Delta t = \frac{s}{\beta c} + \Delta t. \tag{7.64}$$

If the beam is held at constant energy, then we may make a canonical transformation of the time coordinate Δt to rf phase φ given by

$$\varphi = \omega_{rf} \Delta t. \tag{7.65}$$

If acceleration is assumed to be adiabatically slow, so that ω_{rf} changes very slowly, and the magnetic guide fields track the momentum of the synchronous particle, keeping the synchronous particle on a fixed trajectory, we can allow for an adiabatic energy ramp according to

$$U_{sy} = U_0 + \frac{qV \sin \phi_0}{L} s, \tag{7.66}$$

where the energy gain per turn $[qV \sin \phi_0]$ is much less than the total energy U_s. In this case it might not unreasonable to use φ as the longitudinal coordinate, so long as we are prepared to allow for adiabatic damping of the phase space areas. To convert the time coordinate into an rf phase angle relative to the phase of the synchronous particle, we can use the generating function

$$F_2(x, p_x, t, W; s) = x p_x + \left[\omega_{\mathrm{rf}} W - \left(U_0 + \frac{qV \sin\phi_0}{L} s \right) \right] t$$
$$- \frac{2\pi h}{L} W s + \frac{qV \pi h \sin\phi_0}{\omega_{\mathrm{rf}} L^2} s^2, \qquad (7.67)$$

to find a new canonical momentum W corresponding to the phase coordinate. This is what was used to arrive at Eq. (CM: 7.61)[1]

Before proceeding down this path it will behoove us to examine the effect of ramping the energy. The deviation in energy of another particle of energy U from the synchronous particle may be defined as

$$\Delta U = U - U_{\mathrm{sy}}. \qquad (7.68)$$

For the synchronous particle the phase of the rf cavity should be

$$\phi_{\mathrm{sy}} = \phi_0 + \int_0^{t_{\mathrm{sy}}} \omega_{\mathrm{rf}} \, dt$$
$$= \phi_0 + \int_0^{t_{\mathrm{sy}}} \frac{2\pi h \beta c}{L} \, dt_{\mathrm{sy}} \qquad (7.69)$$

With changing energy and the velocity dependence of ω_{rf}, calculation of this integral becomes a problem and φ does not appear to be such an attractive candidate for a canonical coordinate. This is why Chris Iselin chose to take $\zeta = -c\Delta t$ as the longitudinal coordinate variable in the MAD program[33]. Of course there are other parameters which are not necessarily constants in real accelerators. It is quite common to vary the radial position of the closed orbit, as well as the synchronous phase of the rf – particularly during the phase jump at transition crossing. Pulsed quadrupoles are frequently used to cause a rapid change in the transition energy at transition during acceleration.

If we consider a ramp with a constant increase of energy per turn

$$U_{\mathrm{sy}} = U_0 + Rs, \quad \text{with} \quad R = \frac{qV}{L} \sin\phi_0, \qquad (7.70)$$

then the time evolution as a function of path length of the synchronous

---

[1]The formalism of Suzuki[65] was followed in writing CM: § 7.6.

particles is given by

$$
\begin{aligned}
t_{\mathrm{sy}}(s) &= \int_0^s \frac{ds}{\beta c} \\
&= \int_0^s \left[1 - \left(\frac{mc^2}{U_0 + Rs'} \right)^2 \right]^{1/2} ds' \\
&= \frac{mc^2}{R} \int_{\frac{U_0}{mc^2}}^{\frac{U_0 + Rs}{mc^2}} \sqrt{1 - \xi^{-2}}\, d\xi, &\qquad s' &= \frac{\xi mc^2 - U_0}{R} \\
&= -\frac{mc^2}{R} \int_{\frac{mc^2}{U_0}}^{\frac{mc^2}{U_0 + Rs}} \sqrt{1 - \eta^2}\, \frac{d\eta}{\eta^2}, &\qquad \xi &= \frac{1}{\eta} \\
&= \frac{mc^2}{R} \int_{\cos^{-1}\left(\frac{mc^2}{U_0} \right)}^{\cos^{-1}\left(\frac{mc^2}{U_0 + Rs} \right)} \tan^2 \theta\, d\theta, &\qquad \eta &= \cos\theta \\
&= \frac{mc^2}{R} \left[\tan\theta - \theta \right] \Bigg|_{\cos^{-1}\left(\frac{mc^2}{U_0} \right)}^{\cos^{-1}\left(\frac{mc^2}{U_0 + Rs} \right)} \\
&= \frac{mc^2}{2R} \left[\frac{U_0 + Rs}{mc^2} \sqrt{1 - \left(\frac{mc^2}{U_0 + Rs} \right)^2} - \frac{U_0}{mc^2} \sqrt{1 - \left(\frac{mc^2}{U_0} \right)^2} \right. \\
&\qquad \left. + \cos^{-1}\left(\frac{mc^2}{U_0 + Rs} \right) - \cos^{-1}\left(\frac{mc^2}{U_0} \right) \right] \\
&= \frac{mc^2}{2R} \left[\beta\gamma - \beta_0\gamma_0 + \cos^{-1}\left(\frac{1}{\gamma} \right) - \cos^{-1}\left(\frac{1}{\gamma_0} \right) \right] &\qquad (7.71)
\end{aligned}
$$

Provided that the ramping is sufficiently slow, then acceleration may be treated adiabatically.

At least in the adiabatic case, we can find the new canonical coordinate and Hamiltonian from Eq. (7.67):

$$
-U = \frac{\partial F_2}{\partial t} = \omega_{\mathrm{rf}} W - \left(U_0 + \frac{qV \sin\phi_0}{L} s \right), \qquad (7.72)
$$

$$
\varphi = \frac{\partial F_2}{\partial W} = \omega_{\mathrm{rf}} t - \frac{2\pi h}{L} s, \qquad (7.73)
$$

with the derivative of the generating function:

$$
\begin{aligned}
\frac{\partial F_2}{\partial s} &= -\frac{qV \sin \phi_0}{L} t - \frac{2\pi h}{L} s + \frac{2\pi h qV \sin \phi_0}{\omega_{\rm rf} L^2} s \\
&= -\frac{qV \sin \phi_0}{L} \left[\frac{2\pi h}{\omega_{\rm rf} L} s + \Delta t \right] - \frac{2\pi h}{L} s + \frac{2\pi h qV \sin \phi_0}{\omega_{\rm rf} L^2} s \\
&= -\frac{qV \sin \phi_0}{L} \frac{\varphi}{\omega_{\rm rf}} - \frac{2\pi h}{L} s \\
&= -\frac{qV \sin \phi_0}{L} \frac{\varphi}{\omega_{\rm rf}} - \frac{s}{\lambda_{\rm rf}^*},
\end{aligned} \tag{7.74}
$$

where we have written $\lambda_{\rm rf}^* = L/2\pi h$ as a constant effective rf wavelength corrected for the velocity.[2] Ignoring the vertical motion, the new Hamiltonian is

$$
\begin{aligned}
H_1(x, p_x, \varphi, W; s) &= \mathrm{H} + \frac{\partial F_2}{\partial s} \\
&= \frac{p_{\rm sy}}{\rho} x + \frac{p_{\rm sy} K}{2} x^2 + \frac{qV}{\omega_{\rm rf}} \sum_{j=-\infty}^{\infty} \delta(s - jL) \cos\left(\phi_0 + \varphi + \frac{s}{\lambda_{\rm rf}^*} \right) \\
&\quad - \left(1 + \frac{x}{\rho} \right) \left[\frac{(U_{\rm sy} - \omega_{\rm rf} W)^2 - m^2 c^4}{c^2} - p_x^2 \right]^{1/2} - \frac{qV \sin \phi_0}{L} \frac{\varphi}{\omega_{\rm rf}} - \frac{s}{\lambda_{\rm rf}^*} \\
&= \frac{p_{\rm sy}}{\rho} x + \frac{p_{\rm sy} K}{2} x^2 + \frac{qV}{\omega_{\rm rf}} \sum_{j=-\infty}^{\infty} \delta(s - jL) \cos\left(\phi_0 + \varphi + \frac{s}{\lambda_{\rm rf}^*} \right) \\
&\quad - \left(1 + \frac{x}{\rho} \right) \left[p_{\rm sy}^2 - \frac{2\omega_{\rm rf} U_{\rm sy}}{c^2} W + \frac{\omega_{\rm rf}^2}{c^2} W^2 - p_x^2 \right]^{1/2} \\
&\quad - \frac{qV \sin \phi_0}{L} \frac{\varphi}{\omega_{\rm rf}} - \frac{s}{\lambda_{\rm rf}^*}.
\end{aligned} \tag{7.75}
$$

---

[2] The true rf wavelength is actually $\lambda_{\rm rf} = L/2\pi h\beta$ and is a function of the momentum of the particle, whereas $\lambda_{\rm rf}^*$ is a constant depending only on the design circumference and harmonic number.

With a few approximations and a bit more algebra this becomes

H_1

$$\simeq \frac{p_{sy}}{\rho} x + \frac{p_{sy}K}{2} x^2 + \frac{qV}{\omega_{rf}} \sum_{j=-\infty}^{\infty} \delta(s - jL) \cos\left(\phi_0 + \varphi + \frac{s}{\lambda_{rf}^*}\right)$$

$$- p_{sy}\left(1 + \frac{x}{\rho}\right)\left[1 - \frac{\omega_{rf}U_{sy}}{p_{sy}^2 c^2} W + \left(\frac{\omega_{rf}^2}{2p_{sy}^2 c^2} - \frac{1}{8}\frac{4U_{sy}^2\omega_{rf}^2}{p_{sy}^4 c^4}\right) W^2 - \frac{1}{2}\frac{p_x^2}{p_{sy}^2}\right]$$

$$- \frac{qV\sin\phi_0}{L}\frac{\varphi}{\omega_{rf}} - \frac{s}{\lambda_{rf}^*}$$

$$\simeq p_{sy}\left\{-1 + \frac{K}{2} x^2 + \frac{qV}{\omega_{rf}p_{sy}} \sum_{j=-\infty}^{\infty} \delta(s - jL) \cos\left(\phi_0 + \varphi + \frac{s}{\lambda_{rf}^*}\right)\right.$$

$$+ \left(1 + \frac{x}{\rho}\right)\left[\frac{\omega_{rf}U_{sy}}{p_{sy}^2 c^2} W + \frac{m^2\omega_{rf}^2}{2p_{sy}^4} W^2 + \frac{1}{2}\frac{p_x^2}{p_{sy}^2}\right] - \frac{qV\sin\phi_0}{Lp_{sy}}\frac{\varphi}{\omega_{rf}} - \frac{s}{\lambda_{rf}^*}\right\}$$

$$\simeq p_{sy}\left\{-1 + \frac{K}{2} x^2 + \frac{qV}{\omega_{rf}p_{sy}} \sum_{j=-\infty}^{\infty} \delta(s - jL) \cos\left(\phi_0 + \varphi + \frac{s}{\lambda_{rf}^*}\right)\right.$$

$$+ \left(1 + \frac{x}{\rho}\right)\left[\frac{1}{\lambda_{rf}^*}\frac{W}{p_{sy}} + \frac{1}{\gamma^2\lambda_{rf}^{*2}}\left(\frac{W}{p_{sy}}\right)^2 + \frac{1}{2}\left(\frac{p_x}{p_{sy}}\right)^2\right]$$

$$\left. - \frac{qV\sin\phi_0}{Lp_{sy}}\frac{\varphi}{\omega_{rf}} - \frac{s}{\lambda_{rf}^*}\right\}. \tag{7.76}$$

This is essentially the same as Eq. (CM: 7.61) but with the longitudinal variables written as

$$\varphi = \omega_{rf}\,\Delta t, \quad \text{and} \quad W = -\frac{\Delta U}{\omega_{rf}}. \tag{7.77}$$

Here L is the circumference, and we have assumed magnets with only transverse fields and no horizontal-vertical coupling:

$$K = \frac{1}{\rho^2} + \frac{q}{p}\left(\frac{\partial B_y}{\partial x}\right)_0. \tag{7.78}$$

If we want to calculate matrices for the basic magnetic elements, i. e., normal quads and dipoles, then the summation drops out, since $\delta(s - jL) = 0$ and $V = 0$ away from the rf cavities. So keeping only terms to second order in the canonical variables, we have

$$H_1 \simeq -p_s + \frac{p_s K}{2}x^2 + \frac{p_x^2}{2p_s}$$
$$+ \left(\frac{U_{\text{sy}}\omega_{\text{rf}}}{p_{\text{sy}}c^2} - \frac{2\pi h}{L}\right)W + \frac{m^2\omega_{\text{rf}}^2}{2p_{\text{sy}}^3}W^2 + \frac{U_{\text{sy}}\omega_{\text{rf}}}{\rho p_{\text{sy}}c^2}Wx$$
$$\simeq -p_{\text{sy}} + \frac{p_{\text{sy}}K}{2}x^2 + \frac{p_x^2}{2p_{\text{sy}}} + \frac{m^2\omega_{\text{rf}}^2}{2p_{\text{sy}}^3}W^2 + \frac{U_{\text{sy}}\omega_{\text{rf}}}{\rho p_{\text{sy}}c^2}Wx, \quad (7.79)$$

since the two terms in the coefficient of W cancel. We may rescale the Hamiltonian by $1/p_{\text{sy}}$ getting

$$H_{1.5} \simeq -1 + \frac{K}{2}x^2 + \frac{1}{2}w_x^2 + \frac{1}{\gamma^2 \lambdabar_{\text{rf}}^{*2}}w_\varphi^2 + \frac{1}{\rho\lambdabar_{\text{rf}}^*}w_\varphi x, \quad (7.80)$$

with the new canonical momenta

$$w_x = \frac{p_x}{p_{\text{sy}}}, \quad \text{and} \quad (7.81)$$

$$w_\varphi = \frac{W}{p_{\text{sy}}} = -\frac{\Delta U}{\omega_{\text{rf}}p_{\text{sy}}} = -\frac{\frac{\beta\gamma mc^3}{\gamma mc^2}}{\frac{2\pi h\beta c}{L}}\frac{\Delta p}{p_{\text{sy}}}$$

$$= -\lambdabar_{\text{rf}}^*\frac{\Delta p}{p_{\text{sy}}}. \quad (7.82)$$

In this case with φ and w_ϕ as canonically conjugate, the longitudinal emittance would have units of length (meters), just as for the horizontal and vertical planes. (Of course this should be obvious since then all three emittances come from the common Hamiltonian $H_{1.5}$.) In the paraxial approximation, we obviously have $w_x \simeq x'$.

7.4.1 *Variations of the longitudinal canonical variables*

There are several different combinations for the longitudinal canonical variables, for example:

$$(z, \delta) = \left(z, \frac{\Delta p}{p_0}\right), \quad (7.83)$$

$$\left(-c\Delta t, \frac{\Delta U}{p_0 c}\right), \quad (7.84)$$

$$\left(-\frac{(t-t_0)v_0\gamma_0}{\gamma_0+1}, \frac{K-K_0}{K_0}\right), \quad (7.85)$$

$$(\varphi, W), \tag{7.86}$$

$$(\varphi, w_\phi), \tag{7.87}$$

The two pairs in Eqs. (7.86 and 7.82) were previously discussed in § 7.4 with Eqs. (7.77 and 7.82).

Conversion from the pair in Eq. (7.83) to the pair in Eq. (7.84) used in the MAD[3,4] program may be accomplished by the simple transformation

$$\begin{pmatrix} z \\ \frac{\Delta p}{p_0} \end{pmatrix} = \begin{pmatrix} -\beta_0 c(t - t_0) \\ \frac{\Delta p}{p_0} \end{pmatrix} = \begin{pmatrix} -\beta_0 c \Delta t) \\ \beta_0^{-1} \frac{\Delta u}{p_0 c} \end{pmatrix} = \begin{pmatrix} \beta_0 & 0 \\ 0 & \beta_0^{-1} \end{pmatrix} \begin{pmatrix} -c\Delta t \\ \frac{\Delta u}{p_0 c} \end{pmatrix} \tag{7.88}$$

The usual definition of dispersion defined as the particular solution to the inhomogeneous horizontal Hill's equation in Eq. (CM: 5.77) must be modified by a factor of β_0 to agree with the value calculated by MAD.

The pair in Eq. (7.85) are used in the program COSY Infinity[5] which may be obtained as follows. If we start from the good canonical pair:

$$(\Delta t, -\Delta U) \tag{7.89}$$

and rescale as usual by dividing by p_0, we may write the phase-space area element as

$$\Delta t \left(-\frac{\Delta U}{p_0} \right) = \frac{c\Delta t \, \Delta K}{p_0 c}, \tag{7.90}$$

where $\Delta K = \Delta U$ is the change in kinetic energy $K = (\gamma - 1)mc^2$. Writing $p_0 c$ in terms of the kinetic energy, we have

$$\begin{aligned} p_0 c &= mc^2 \gamma_0 \beta_0 = mc^2 (\gamma_0 - 1)\sqrt{\frac{\gamma_0 + 1}{\gamma_0 - 1}} = K_0 \sqrt{\frac{\gamma_0 + 1}{\gamma_0 - 1}} \\ &= \frac{K_0(\gamma_0 + 1)}{\sqrt{\gamma_0^2 - 1}} = \frac{K_0(\gamma_0 + 1)}{\gamma_0 \beta_0}. \end{aligned} \tag{7.91}$$

Substituting this into Eq. (7.90) we find

$$\Delta t \left(-\frac{\Delta U}{p_0} \right) = -\frac{\Delta t \, \gamma_0 v_0}{\gamma_0 + 1} \frac{\Delta K}{K_0}, \tag{7.92}$$

and we see that Eq. (7.85) must also be a good canonical pair that preserves the area elements of longitudinal phase space.

7.4.2   *Transport matrices for a few elements*

Using the Lie algebraic method demonstrated in CM: § 3.7, matrices for a drift, quadrupole, and sector bend may obtained:

$$
\mathbf{M}_{\text{drift}} = \begin{pmatrix} 1 & l & 0 & 0 \\ 0 & 1 & 0 & 0 \\ 0 & 0 & 1 & \frac{l}{\gamma^2 \lambda_{\text{rf}}^{*\,2}} \\ 0 & 0 & 0 & 1 \end{pmatrix}, \tag{7.93}
$$

$$
\mathbf{M}_{\text{quad}} = \begin{pmatrix} \cos(\sqrt{k}l) & \frac{1}{\sqrt{k}}\sin(\sqrt{k}l) & 0 & 0 \\ -\sqrt{k}\sin(\sqrt{k}l) & \cos(\sqrt{k}l) & 0 & 0 \\ 0 & 0 & 1 & \frac{l}{\gamma^2 \lambda_{\text{rf}}^{*\,2}} \\ 0 & 0 & 0 & 1 \end{pmatrix}, \tag{7.94}
$$

$\mathbf{M}_{\text{sbend}} =$

$$
\begin{pmatrix} \cos\left[\sqrt{1-n}\,\theta\right] & \frac{\rho\sin\left[\sqrt{1-n}\,\theta\right]}{\sqrt{1-n}} & 0 & \frac{\rho\left(1-\cos\left[\sqrt{1-n}\,\theta\right]\right)}{(1-n)\lambda_{\text{rf}}^{*}} \\ \frac{\sqrt{1-n}\,\sin\left[\sqrt{1-n}\,\theta\right]}{\rho} & \cos\left[\sqrt{1-n}\,\theta\right] & 0 & \frac{\sin\left[\sqrt{1-n}\,\theta\right]}{\sqrt{1-n}\lambda_{\text{rf}}^{*}} \\ -\frac{\sin\left[\sqrt{1-n}\,\theta\right]}{\sqrt{1-n}\lambda_{\text{rf}}^{*}} & -\frac{\rho\left(1-\cos\left[\sqrt{1-n}\,\theta\right]\right)}{(1-n)\lambda_{\text{rf}}^{*}} & 1 & \frac{\rho}{\lambda_{\text{rf}}^{*\,2}}\left\{\frac{\theta}{\gamma^2} - \frac{\sqrt{1-n}\,\theta-\sin\left[\sqrt{1-n}\,\theta\right]}{(1-n)^{-3/2}}\right\} \\ 0 & 0 & 0 & 1 \end{pmatrix},
$$

$$\tag{7.95}$$

where we have used $H_{1.5}$ with the canonical pairs (x, w_x) and (ϕ, w_ϕ) for the radial and longitudinal canonical variables. To write these matrices with respect to the (z, δ) system, we leave it as an exercise to show, in effect, that we may simply set $\lambda_{\text{rf}}^{*} = 1$.

Chapter 8

Problems of Chapter 8: Synchrotron Radiation

8.1 Problem 8–1: $\mathcal{D}$ for a simple lattice

For a separated function ring with identical dipole magnets of bending radius, ρ, show that

$$\mathcal{D} = \frac{L\alpha_p}{2\pi\rho}, \tag{8.1}$$

where L is the circumference, and α_p is the momentum compaction.

Solution:

Let us elaborate Eq. (CM: 8.35) as follows:

$$\mathcal{D} \simeq \frac{1}{cU_\gamma} \oint P_\gamma \frac{\eta(s)}{\rho} \, ds \tag{CM: 8.35}$$

$$= \frac{P_\gamma}{cU_\gamma} \oint \frac{\eta(s)}{\rho} \, ds = \frac{P_\gamma}{cU_\gamma} L\alpha_p. \tag{8.2}$$

Notice that P_γ and U_γ are respectively the power and the energy pertaining to the synchrotron radiation emitted over a revolution; therefore we have

$$P_\gamma = \frac{U_\gamma}{\tau_{\text{mag}}} = \frac{cU_\gamma}{2\pi\rho} \qquad \Longrightarrow \qquad \frac{P_\gamma}{cU_\gamma} = \frac{1}{2\pi\rho} \tag{8.3}$$

since the radiation emission takes place inside the bending magnets only, and moreover, the whole synchrotron physics deals with ultrarelativistic particles. Hence, inserting Eq. (8.3) into Eq. (8.2), we obtain just

$$\mathcal{D} = \frac{L\alpha_p}{2\pi\rho}. \tag{8.4}$$

8.2   Problem 8–2: Damping time for 6-d volume

Show that the damping time for the six dimensional phase space volume of
a beam is just $\tau_0/8 = U_s/(4U_\gamma)$.

Solution:

The phase space volume is proportional to the product of the phase space
coordinates averaged over the full particle distribution:

$$V \propto \langle xx' \, yy' zu \rangle$$
$$\simeq \langle A_x e^{-t/\tau_x} A_{x'} e^{-t/\tau_x} \rangle \langle A_y e^{-t/\tau_y} A_{y'} e^{-t/\tau_y} \rangle \langle A_z e^{-t/\tau_u} A_u e^{-t/\tau_u} \rangle$$
$$\simeq (\pi \epsilon_{\mathrm{H,rms}} \, \pi \epsilon_{\mathrm{V,rms}} \, \pi \epsilon_{\mathrm{u,rms}}) e^{-t/\tau}, \qquad (8.5)$$

where the damping rate is

$$\frac{1}{\tau} = \frac{2}{\tau_x} + \frac{2}{\tau_y} + \frac{2}{\tau_u}. \qquad (8.6)$$

Recalling Robinson's damping partition theorem:

$$\frac{1}{\tau_x} + \frac{1}{\tau_y} + \frac{1}{\tau_u} = \frac{4}{\tau_0}, \qquad (\mathrm{CM: \ 8.80})$$

Eq. (8.6) becomes

$$\frac{1}{\tau} = \frac{8}{\tau_0} = \frac{4U_\gamma}{U_s}, \quad \text{or equivalently} \quad \tau = \frac{\tau_0}{8} = \frac{U_s}{4U_\gamma}, \qquad (8.7)$$

since

$$\tau_0 = \frac{2U_s}{U_\gamma} \tau_s. \qquad (\mathrm{CM: \ 8.77})$$

Caveat:

Notice that this result is almost exact if we consider small amplitude synchrotron oscillations only. Nevertheless, it is obvious that the square-power
effect cannot disappear, as a glance at the integral in

$$I_L = \frac{h^2 \eta_{\mathrm{tr}} \omega_s^2}{\beta^2 \gamma mc^2} \oint W^2 \, dt \qquad (\mathrm{CM: \ 7.87})$$

clearly suggests.

8.3 Problem 8–3: Double-bend achromat

An achromatic bend (the double-bend achromat) may be made from two dipoles with a horizontally focusing quadrupole between them. The transfer matrix through the achromat is of the form:

$$\mathbf{M} = \mathbf{B}(\theta)\mathbf{L}\left(\tfrac{1}{2}\mathbf{Q}\right)\left(\tfrac{1}{2}\mathbf{Q}\right)\mathbf{LB}(\theta). \tag{8.8}$$

a) Use thin the lens approximation for quads and small angle approximation for bends to find the dispersion in the middle of the quad. Write the focal length in terms of the drift length and bend parameters.

$$\text{Hint:} \qquad \begin{pmatrix} \eta_c \\ 0 \\ 1 \end{pmatrix} = \left(\tfrac{1}{2}\mathbf{Q}\right)\mathbf{LB}\begin{pmatrix} 0 \\ 0 \\ 1 \end{pmatrix}. \tag{8.9}$$

b) Show that the dispersion is again zero ($\eta = \eta' = 0$) after the bend.

Solution:

This sequence of elements is the basis of the Chasmann-Green lattice[15] used in synchrotron light sources with low emittance beams.

Even though the problem says to use the thin lens and small angle bend approximations, we will give the more complicated solution with the "thick" matrices.

a) We only need to consider 3×3 matrices for the coordinates x, x', and δ. Define the elements

$$\mathbf{B} = \begin{pmatrix} \cos\theta & \rho\sin\theta & \rho(1-\cos\theta) \\ -\tfrac{1}{\rho}\sin\theta & \cos\theta & \sin\theta \\ 0 & 0 & 1 \end{pmatrix}, \tag{8.10}$$

$$\mathbf{L} = \begin{pmatrix} 1 & \lambda & 0 \\ 0 & 1 & 0 \\ 0 & 0 & 1 \end{pmatrix}, \tag{8.11}$$

$$\mathbf{Q} = \begin{pmatrix} \cos\phi & \tfrac{1}{\sqrt{k}}\sin\phi & 0 \\ -\sqrt{k}\sin\phi & \cos\phi & 0 \\ 0 & 0 & 1 \end{pmatrix}, \qquad \text{where} \quad \phi = \sqrt{k}\,l, \tag{8.12}$$

$$\hat{\mathbf{Q}} = \left(\tfrac{1}{2}\mathbf{Q}\right) = \begin{pmatrix} \cos\tfrac{\phi}{2} & \tfrac{1}{\sqrt{k}}\sin\tfrac{\phi}{2} & 0 \\ -\sqrt{k}\sin\tfrac{\phi}{2} & \cos\tfrac{\phi}{2} & 0 \\ 0 & 0 & 1 \end{pmatrix}. \tag{8.13}$$

It is useful to split the achromat into pieces with mirror symmetry:

$$\mathbf{M} = \begin{pmatrix} C & S & D \\ C' & S' & D' \\ 0 & 0 & 1 \end{pmatrix} = \mathbf{B\,L\,Q\,L\,B} = \mathbf{M_2\,M_1}$$

(8.14)

with

$$\mathbf{M_1} = \widehat{\mathbf{Q}}\,\mathbf{L}\,\mathbf{B}, \quad \text{and} \quad \mathbf{M_2} = \mathbf{B}\,\mathbf{L}\,\widehat{\mathbf{Q}}. \tag{8.15}$$

For an achromat we then want

$$\mathbf{M_1} \begin{pmatrix} 0 \\ 0 \\ 1 \end{pmatrix} = \begin{pmatrix} \eta_c \\ 0 \\ 1 \end{pmatrix}, \tag{8.16}$$

where η_c is the dispersion in the center of the quadrupole magnet and $\eta_c' = 0$.

$$\begin{pmatrix} \eta_c \\ \eta_0' \\ 1 \end{pmatrix} = \mathbf{M_1} \begin{pmatrix} 0 \\ 0 \\ 1 \end{pmatrix}$$

$$= \begin{pmatrix} \cos\frac{\phi}{2} & \frac{1}{\sqrt{k}}\sin\frac{\phi}{2} & 0 \\ -\sqrt{k}\sin\frac{\phi}{2} & \cos\frac{\phi}{2} & 0 \\ 0 & 0 & 1 \end{pmatrix} \begin{pmatrix} 1 & \lambda & 0 \\ 0 & 1 & 0 \\ 0 & 0 & 1 \end{pmatrix}$$

$$\begin{pmatrix} \cos\theta & \rho\sin\theta & \rho(1-\cos\theta) \\ -\frac{1}{\rho}\sin\theta & \cos\theta & \sin\theta \\ 0 & 0 & 1 \end{pmatrix} \begin{pmatrix} 0 \\ 0 \\ 1 \end{pmatrix}; \tag{8.17}$$

$$\begin{pmatrix} \eta_c \\ \eta_c' \\ 1 \end{pmatrix} = \begin{pmatrix} \cos\frac{\phi}{2} & \frac{1}{\sqrt{k}}\sin\frac{\phi}{2} + \lambda\cos\frac{\phi}{2} & 0 \\ -\sqrt{k}\sin\frac{\phi}{2} & \cos\frac{\phi}{2} - \lambda\sqrt{k}\sin\frac{\phi}{2} & 0 \\ 0 & 0 & 1 \end{pmatrix} \begin{pmatrix} \rho(1-\cos\theta) \\ \sin\theta \\ 1 \end{pmatrix}$$

$$= \begin{pmatrix} \rho(1-\cos\theta)\cos\frac{\phi}{2} + \sin\theta\left[\frac{1}{\sqrt{k}}\sin\frac{\phi}{2} + \lambda\cos\frac{\phi}{2}\right] \\ -\rho(1-\cos\theta)\sqrt{k}\sin\frac{\phi}{2} + \sin\theta\left[\cos\frac{\phi}{2} - \lambda\sqrt{k}\sin\frac{\phi}{2}\right] \\ 1 \end{pmatrix}$$

$$= \begin{pmatrix} [\rho(1-\cos\theta) + \lambda\sin\theta]\cos\frac{\phi}{2} + \frac{1}{\sqrt{k}}\sin\theta\sin\frac{\phi}{2} \\ -\sqrt{k}\,[\rho(1-\cos\theta) + \lambda\sin\theta]\sin\frac{\phi}{2} + \sin\theta\cos\frac{\phi}{2} \\ 1 \end{pmatrix}. \tag{8.18}$$

For $\eta_c' = 0$ the second row gives

$$\rho(1 - \cos\theta) + \lambda\sin\theta = \frac{1}{\sqrt{k}}\sin\theta\cot\frac{\phi}{2}. \tag{8.19}$$

Solve this for λ to get

$$\lambda = \frac{1}{\sqrt{k}}\cot\frac{\phi}{2} - \rho\frac{2\sin^2\frac{\theta}{2}}{2\sin\frac{\theta}{2}\cos\frac{\theta}{2}} = \frac{1}{\sqrt{k}}\cot\frac{\phi}{2} - \rho\tan\frac{\theta}{2}. \tag{8.20}$$

Combining Eq. (8.19) with the first row of Eq. (8.18) yields

$$\eta_c = \frac{1}{\sqrt{k}}\sin\theta\cot\frac{\phi}{2}\cos\frac{\phi}{2} + \frac{1}{\sqrt{k}}\sin\theta\sin\frac{\phi}{2}$$

$$= \frac{\sin\theta}{\sqrt{k}\sin\frac{\phi}{2}}. \tag{8.21}$$

With the small-angle, and thin-lens approximations this becomes

$$\eta_c \simeq \frac{2\theta}{kl} \simeq 2\theta f. \tag{8.22}$$

for a horizontally focusing quadrupole of focal length $f = 1/\left(\sqrt{k}\,l\right)$.

b) Propagating from the middle to the end of the achromat,

$$\begin{pmatrix} \eta_e \\ \eta_e' \\ 1 \end{pmatrix} = \mathbf{M_2}\begin{pmatrix} \eta_c \\ 0 \\ 1 \end{pmatrix} = \mathbf{B}\,\mathbf{L}\,\widehat{\mathbf{Q}}\begin{pmatrix} \eta_c \\ 0 \\ 1 \end{pmatrix}$$

$$= \begin{pmatrix} \cos\theta & \rho\sin\theta & \rho(1-\cos\theta) \\ -\frac{1}{\rho}\sin\theta & \cos\theta & \sin\theta \\ 0 & 0 & 1 \end{pmatrix}\begin{pmatrix} 1 & \lambda & 0 \\ 0 & 1 & 0 \\ 0 & 0 & 1 \end{pmatrix}$$

$$\times \begin{pmatrix} \cos\frac{\phi}{2} & \frac{1}{\sqrt{k}}\sin\frac{\phi}{2} & 0 \\ -\sqrt{k}\sin\frac{\phi}{2} & \cos\frac{\phi}{2} & 0 \\ 0 & 0 & 1 \end{pmatrix}\begin{pmatrix} \eta_c \\ 0 \\ 1 \end{pmatrix}$$

$$= \begin{pmatrix} \cos\theta & \rho\sin\theta & \rho(1-\cos\theta) \\ -\frac{1}{\rho}\sin\theta & \cos\theta & \sin\theta \\ 0 & 0 & 1 \end{pmatrix}\begin{pmatrix} 1 & \lambda & 0 \\ 0 & 1 & 0 \\ 0 & 0 & 1 \end{pmatrix}\begin{pmatrix} \eta_c\cos\frac{\phi}{2} \\ -\eta_c\sqrt{k}\sin\frac{\phi}{2} \\ 1 \end{pmatrix}$$

$$= \begin{pmatrix} \cos\theta & \rho\sin\theta & \rho(1-\cos\theta) \\ -\frac{1}{\rho}\sin\theta & \cos\theta & \sin\theta \\ 0 & 0 & 1 \end{pmatrix}\begin{pmatrix} \eta_c(\cos\frac{\phi}{2} - \lambda\sqrt{k}\sin\frac{\phi}{2}) \\ -\eta_c\sqrt{k}\sin\frac{\phi}{2} \\ 1 \end{pmatrix}. \tag{8.23}$$

with

$$
\eta_e = \eta_c \left[\cos\theta \cos\frac{\phi}{2} - \left(\frac{1}{\sqrt{k}} \cot\frac{\phi}{2} - \rho\tan\frac{\theta}{2} \right) \sqrt{k} \cos\theta \sin\frac{\phi}{2} \right.
$$
$$
\left. - \rho\sqrt{k} \sin\theta \sin\frac{\phi}{2} \right] + \rho(1 - \cos\theta)
$$
$$
= \eta_c \, \rho\sqrt{k} \sin\frac{\phi}{2} \left(\tan\frac{\theta}{2} \cos\theta - \sin\theta \right) + \rho(1 - \cos\theta)
$$
$$
= \eta_c \, \rho\sqrt{k} \sin\frac{\phi}{2} \left[\tan\frac{\theta}{2} \left(2\cos^2\frac{\theta}{2} - 1 \right) - \sin\theta \right] + \rho(1 - \cos\theta)
$$
$$
= -\rho \left(\frac{\sin\theta}{\sqrt{k}\sin\frac{\phi}{2}} \sqrt{k} \sin\frac{\phi}{2} \tan\frac{\theta}{2} \right) + 2\rho\sin^2\frac{\theta}{2}
$$
$$
= \rho \left(-2\sin^2\frac{\theta}{2} + 2\sin^2\frac{\theta}{2} \right)
$$
$$
= 0, \tag{8.24}
$$

and

$$
\eta_e' = \eta_c \left[-\frac{1}{\rho} \sin\theta \cos\frac{\phi}{2} + \frac{1}{\rho} \sin\theta \left(\frac{1}{\sqrt{k}} \cot\frac{\phi}{2} - \rho\tan\frac{\theta}{2} \right) \sqrt{k} \sin\frac{\phi}{2} \right.
$$
$$
\left. - \sqrt{k} \cos\theta \sin\frac{\phi}{2} \right] + \sin\theta
$$
$$
= -\eta_c \sqrt{k} \sin\frac{\phi}{2} \left[\sin\theta \tan\frac{\theta}{2} + \cos\theta \right] + \sin\theta
$$
$$
= -\frac{\sin\theta}{\sqrt{k}\sin\frac{\phi}{2}} \sqrt{k} \sin\frac{\phi}{2} \left[2\sin\frac{\theta}{2} \cos\frac{\theta}{2} \frac{\sin\frac{\theta}{2}}{\cos\frac{\theta}{2}} + \cos^2\frac{\theta}{2} - \sin^2\frac{\theta}{2} \right] + \sin\theta
$$
$$
= -\sin\theta \left[\sin^2\frac{\theta}{2} + \cos^2\frac{\theta}{2} \right] + \sin\theta
$$
$$
= 0. \tag{8.25}
$$

8.4 Problem 8–4: Light source ring damping times

A light source ring has eight equal double achromat bends (16 dipoles). Each dipole is 2.7 m long, and the circumference is 176 m. The energy of the beam is 2.5 GeV.

a) Calculate the critical energy of photons radiated in the dipoles.
b) Calculate the total energy lost per turn.
c) Calculate the momentum compaction of the ring.
d) Calculate the damping times τ_x, τ_y, and τ_u.

Solution:

a) Recalling the critical frequency $\omega_c = \frac{3c}{2\rho} \gamma^3$, we have

$$\theta_b = 2\pi/16 = \pi/8. \tag{8.26}$$

$$\rho = 2.7 \ [\text{m}]/(\pi/8) = 6.876 \ [\text{m}], \tag{8.27}$$

$$\gamma = \frac{U_s}{mc^2} = \frac{2.5 \times 10^9 \ [\text{eV}]}{5.11 \times 10^5 \ [\text{eV}]} = 4892, \tag{8.28}$$

$$u_c = \hbar\omega_c = \frac{3\hbar c}{2\rho}\gamma^3$$

$$= \frac{3 \times 197 \times 10^{-9} \ [\text{eV m}]}{2 \times 6.876 \ [\text{m}]} \times 4892^3 = 5030 \ [\text{eV}]. \tag{8.29}$$

b) From Eqs. (CM: 8.14 and CM: 8.15), we have

$$U_\gamma = \frac{C_\gamma U_s^4}{2\pi} \oint \frac{ds}{\rho^2} = \frac{C_\gamma U_s^4}{\rho}$$

$$= 8.85 \times 10^{-5} \left[\frac{\text{m}}{(\text{GeV})^3} \right] \times \frac{(2.5 \ [\text{GeV}])^4}{6.876 [\text{m}]}$$

$$= 0.503 \ [\text{MeV}]. \tag{8.30}$$

c) Recall the formula for momentum compaction:

$$\alpha_p = \frac{1}{L} \oint \frac{\eta(s)}{\rho(s)} \, ds. \tag{CM: 5.98}$$

By symmetry, the integrals over each of the dipoles is the same, so we may write

$$\alpha_p = \frac{16}{L} \int_0^{\rho\theta_b} \frac{\eta(s)}{\rho} \, ds = \frac{16}{L} \int_0^{\theta_b} \eta(\theta) \, d\theta. \tag{8.31}$$

Remember that the double-bend achromat is mirror-symmetric about the center of the quadrupole. Starting from a zero-dispersion straight section, and integrating through one dipole to the middle of an achromat we will have $\eta(\theta) = \rho(1 - \cos\theta)$ inside the dipole, so

$$\alpha_p = \frac{16\rho}{L} \int_0^{\theta_b} (1 - \cos\theta)\, d\theta = \frac{16\rho}{L}(\theta_b - \sin\theta_b)$$

$$= \frac{16 \times 6.876\ [\text{m}]}{176\ [\text{m}]} \left[\frac{\pi}{8} - \sin\frac{\pi}{8}\right] = 0.00626. \tag{8.32}$$

d) The revolution time is

$$\tau_s = \frac{L}{c} = \frac{176\ [\text{s}]}{3 \times 10^8\ [\text{m/s}]} = 0.587\ \mu\text{s}. \tag{8.33}$$

$$\tau_0 = \frac{2U_s}{U_\gamma}\tau_s = 5.84\ [\text{ms}]. \tag{8.34}$$

From Eq. (8.33) we have

$$\mathcal{D} = \frac{1}{cU_\gamma} \oint P_\gamma \eta \frac{1 - 2n}{\rho}\, ds$$

$$= \frac{16}{cU_\gamma} P_\gamma \int_0^{\theta_b} \rho(1 - \cos\theta)\, d\theta, \tag{8.35}$$

since $n = 0$ (no gradient) in a dipole bend. The energy radiated in one full turn is

$$U_\gamma = P_\gamma \frac{2\pi\rho}{c}, \tag{8.36}$$

assuming that the radiation is primarily from the dipoles, since there is no radiation in the drifts and we assume that the closed orbit goes through the center of the quadrupoles. Combining these last two equations gives

$$\mathcal{D} = \frac{16}{2\pi}(\theta_b - \sin\theta_b) = 0.0255. \tag{8.37}$$

From CM: §8.5 we may now write

$$J_x = 1 - \mathcal{D} = 0.9745, \tag{8.38}$$

$$J_y = \mathcal{D} = 1, \tag{8.39}$$

$$J_u = 2 + \mathcal{D} = 2.0255, \tag{8.40}$$

$$\tau_x = \frac{\tau_0}{J_x} = 5.993\ [\text{ms}], \tag{8.41}$$

$$\tau_y = \frac{\tau_0}{J_y} = 5.84\ [\text{ms}], \tag{8.42}$$

$$\tau_u = \frac{\tau_0}{J_u} = 2.883\ [\text{ms}]. \tag{8.43}$$

8.5 Problem 8–5: LHC synchrotron radiation

According to its design, the Large Hadron Collider (LHC) will be capable of accelerating protons to 7 TeV in each of two rings. The circumference will be 26.7 km, and the arc dipole field at 7 TeV will be 8.33 T.

a) Calculate the critical energy of photons.
b) Calculate the energy loss per turn per proton.
c) Calculate the total power radiated by synchrotron radiation for a beam with an average current of 0.56 A.

Solution:

a) The rigidity is

$$\frac{p}{q} = \frac{1}{c}\sqrt{U_s^2 - (mc^2)^2} = 23.35 \times 10^3 \text{ [Tm]}, \tag{8.44}$$

so the bending radius must be

$$\rho = \frac{p}{qB} = 2803 \text{ [m]}. \tag{8.45}$$

Recalling Eq. (CM:8.84), we find the critical energy

$$u_c = \hbar\omega_c = \frac{3}{2}\frac{\hbar c}{\rho}\gamma^3 = 43.8 \text{ [eV]}, \tag{8.46}$$

with $\hbar c \simeq 197$ [MeV fm].

b) From Eq. (CM:8.14) the energy loss per turn is

$$U_\gamma \simeq \frac{C_\gamma U_s^4}{2\pi}\oint\frac{ds}{\rho^2} = \frac{C_\gamma U_s^4}{2\pi}\frac{2\pi\rho}{\rho^2} = \frac{C_\gamma U_s^4}{\rho}$$

$$= \frac{8.85 \times 10^{-5}\left[\frac{m}{(\text{GeV})^3}\right] \times (7 \times 10^3 \text{ [GeV]})^4}{2803 \text{ [m]}}$$

$$= 6.67 \text{ [keV]}. \tag{8.47}$$

c) The average power radiated is

$$\bar{P}_\gamma = U_\gamma\frac{\bar{I}}{e} = 6.67 \left[\frac{\text{keV}}{e}\right] \times 0.56 \text{ [C/s]} = 3.73 \text{ [kW]}. \tag{8.48}$$

Chapter 9

Problems of Chapter 9: RF Linear Accelerators

9.1 Problem 9–1: Acceleration via a plane wave

How strong must the electric field intensity of a traveling plane wave be to accelerate electrons with an energy gradient of 10 MeV/m? (Hint: Use the Poynting vector.)

Solution:

There are perhaps two ways this question could be answered, although the first is rather naive and ignores the hint: Instantaneously an electric field strength of 10 MV/m will accelerate an electron at 10 MV/m over a very short distance; however, the perpendicular magnetic field will cause a twist of the trajectory so that there would not be a long term average acceleration of the electron. Also the electron having a mass would not be traveling with the same velocity as the wave and would fall behind the accelerating crest.

Another way to approach this problem is as Compton scattering from a coherent field of photons with energy transfer from the wave to the electrons. For the plane wave we have the Poynting vector $\vec{S} = \frac{1}{\mu_0}\vec{E} \times \vec{B}$, with $B = E/c$ in a vacuum so that the flow of energy in the wave is

$$S = \frac{E^2}{\mu_0 c}. \tag{9.1}$$

The Poynting vector has units of Wm^{-2}. We want an energy transfer per distance of

$$\frac{dU}{dz} = q\bar{E} = 10 \text{ MeV/m}, \tag{9.2}$$

or for an electron traveling with a velocity v

$$\frac{dU}{dt} = qv\bar{E}. \tag{9.3}$$

Using the Thompson cross section of an electron[50]

$$\sigma_{\mathrm{T}} = \frac{8\pi r_e^2}{3} = 6.65 \times 10^{-29} \ \mathrm{m}^2, \tag{9.4}$$

with r_e being the classical radius of an electron, the pressure of the wave on the electron will apply an average acceleration

$$\frac{dU}{dt} = q\bar{E}v = SA = \frac{E^2 \sigma_{\mathrm{T}}}{\mu_0 c}. \tag{9.5}$$

Solving for E we find

$$E \simeq \sqrt{\frac{\mu_0 c v q \bar{E}}{\sigma_{\mathrm{T}}}} \simeq 1.65 \times 10^{14} \sqrt{\frac{v}{c}} \ \mathrm{V/m}. \tag{9.6}$$

So for a relativistic electron, we would need a plane wave with an electric field of about 170 TV/m.

9.2 Problem 9–2: Power loss in cavity walls

Show that the rf power loss in the conducting walls of a cavity is given by Eq. (CM: 9.16).

$$\langle P_{\text{loss}} \rangle = \frac{R_s}{2} \iint_S |H_\parallel|^2 \, dS. \qquad \text{(CM: 9.16)}$$

Solution:

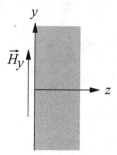

Fig. 9.1 For a time-varying magnetic field intensity component H_y parallel to the cavity wall, there must be a current (flowing into the page) inside the conductor (shaded region) which drops off exponentially with with the depth z.

The component of magnetic field field intensity H_y parallel to the surface of a conductor will have an amplitude $H_y = H_0$ at the surface $z = 0$ (see Fig. 9.1, and from the Helmholtz equation we must have $\nabla^2 H_y = \Gamma^2 H_y$, inside the conductor. By symmetry H_y is only a function of z. Therefore $H_y(z) = H_0 e^{-\Gamma z}$ with $\Gamma = \frac{1+i}{\delta}$. The power loss inside the wall averaged over one cycle will be

$$P_{\text{ave}} = \frac{1}{\tau} \int_0^\tau \int \text{Re}(\vec{J}) \cdot \text{Re}(\vec{E}) \, d^3x \, dt = \int \frac{1}{2} \text{Re}(\vec{J} \cdot \vec{E}^*) \, d^3x$$

$$= \frac{1}{2} \text{Re} \left(\int \sigma \vec{E} \cdot \vec{E}^* \, d^3x \right) = \frac{1}{2} \int \sigma |E|^2 \, d^3x. \qquad (9.7)$$

Solving Ampere's law

$$\nabla \times \vec{H} = (\sigma + i\omega\epsilon)\vec{E} \qquad (9.8)$$

for the electric field inside the conductor, we find

$$\vec{E} = \frac{1}{\sigma + i\omega\epsilon} \nabla \times \vec{H} \simeq \frac{1}{\sigma} \nabla \times \vec{H}, \quad \text{assuming } \omega\epsilon \ll \sigma,$$

$$= \frac{1}{\sigma} \left(-\frac{\partial H_y}{\partial z} \right) \hat{x} = -\frac{1}{\sigma} (-\Gamma H_y) \hat{x}$$

$$= \frac{1+i}{\sigma\delta} H_y \hat{x}. \tag{9.9}$$

Substituting this into the previous equation gives

$$P_{\text{ave}} = \frac{\sigma}{2} \left(\frac{1}{\sigma\delta} \right)^2 \iiint_V |(1+i)|^2 |H_y|^2 \, d^3x$$

$$= \frac{1}{2\sigma\delta^2} \times 2 \iint_S \left(\int_0^\infty H_0^2 e^{-z(1+i)/\delta} e^{-z(1-i)/\delta} \, dz \right) dx \, dy$$

$$= \frac{1}{\sigma\delta^2} \iint_S |H_0|^2 \, dx \, dy \int_0^\infty e^{-2z/\delta} \, dz$$

$$= \frac{1}{2\sigma\delta} \iint_S |H_0|^2 \, dx \, dy$$

$$= \frac{R_s}{2} \iint_S |H_\parallel|^2 \, dS. \tag{9.10}$$

9.3 Problem 9–3: Lowest TM mode of rectangular cavity

Consider a box shaped resonant cavity with a square cross section of width, w, in the transverse directions and length, l, in the longitudinal dimension. Calculate the resonant frequency and Q of the lowest order TM mode.

Solution:

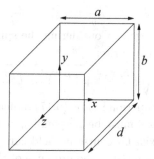

Fig. 9.2   Rectangular box

Consider the shape of the rectangular cavity shown in Fig. 9.2, where the dimensions a, b, d are parallel respectively to $\hat{x}, \hat{y}, \hat{z}$. For all the TM modes we have $B_z = 0$ while E_z is evaluated via the Helmholtz equation

$$\nabla^2 E_z(x,y,z) + k^2 E_z(x,y,z) = 0 \qquad (9.11)$$

Since we are dealing with standing waves, the solution must depend on the coordinates in a separate manner; that is $E_z(x,y,z) = X(x)Y(y)Z(z)$ which, inserted into Eq. (9.11), gives rise to the relation

$$YZ\frac{d^2X}{dx^2} + XZ\frac{d^2Y}{dy^2} + XY\frac{d^2Z}{dz^2} + (k_x^2 + k_y^2 + k_z^2)XYZ, \qquad (9.12)$$

having split the constant k^2 in the sum of other three constants. Hence dividing Eq. (9.12) by XYZ, we obtain

$$\left[\frac{1}{X}\frac{d^2X}{dx^2} + k_x^2\right] + \left[\frac{1}{Y}\frac{d^2Y}{dy^2} + k_y^2\right] + \left[\frac{1}{Z}\frac{d^2Z}{dz^2} + k_z^2\right] = 0. \qquad (9.13)$$

Due to their reciprocal independence, the three expressions within square brackets must be zero yielding thus three equations of the harmonic oscillator type.

$$\frac{d^2X}{dx^2} + k_x^2 X = 0, \qquad \frac{d^2Y}{dy^2} + k_y^2 Y = 0, \qquad \frac{d^2Z}{dz^2} + k_z^2 Z = 0, \qquad (9.14)$$

whose solutions will be sinusoidal functions of arguments $k_x x$, $k_y y$ and $k_z z$, respectively. In order to find their explicit form, we must take into account the boundary conditions:

$$E_z(0, y, z) = E_z(a, y, z) = E_z(x, 0, z) = E_z(x, b, z) = 0, \qquad (9.15)$$

$$E_z(x, y, 0) = E_z(x, y, d) = E_0, \qquad (9.16)$$

where E_0 is the crest electric field. We have

$$E_z = E_0 \sin\left(\frac{m\pi x}{a}\right) \sin\left(\frac{n\pi y}{b}\right) \cos\left(\frac{p\pi z}{d}\right), \qquad (9.17)$$

with

$$\frac{m\pi}{a} = k_x, \qquad \frac{n\pi}{b} = k_y, \qquad \frac{p\pi}{d} = k_z, \qquad (9.18)$$

where m, n and p are integers. Considering the splitting of the constant k shown in Eq. (9.12), we can write:

$$k = \sqrt{\left(\frac{m\pi}{a}\right)^2 + \left(\frac{n\pi}{b}\right)^2 + \left(\frac{p\pi}{d}\right)^2}. \qquad (9.19)$$

The choice of the cosine function in Eq. (9.17) indicates that the component of $\vec{E}$ parallel to the z-axis must be perpendicular to the two cavity walls which encompass this axis. Of course no field can penetrate into the walls parllel to the axis, so this justifies the presence of the two sines. As for the other two components, we shall use Gauss' law $\vec{\nabla} \cdot \vec{E} = \frac{\rho}{\varepsilon_0} = 0$, so

$$\left(\frac{\partial E_x}{\partial x} + \frac{\partial E_y}{\partial y}\right) = -\frac{\partial E_z}{\partial z} = \left(\frac{p\pi}{d}\right) E_0 \sin\left(\frac{m\pi x}{a}\right) \sin\left(\frac{n\pi y}{b}\right) \sin\left(\frac{p\pi z}{d}\right). \qquad (9.20)$$

Proceeding by inspection, we search for two solutions as similar as possible to Eq. (9.17), i. e. a constant times two sines and a cosine in correspondence of each axis:

$$E_x = \frac{\alpha_1 \delta}{K_c^2} \cos\left(\frac{m\pi x}{a}\right) \sin\left(\frac{n\pi y}{b}\right) \sin\left(\frac{p\pi z}{d}\right), \qquad (9.21)$$

$$E_y = \frac{\alpha_2 \delta}{K_c^2} \sin\left(\frac{m\pi x}{a}\right) \cos\left(\frac{n\pi y}{b}\right) \sin\left(\frac{p\pi z}{d}\right). \qquad (9.22)$$

Their partial derivatives are

$$\frac{\partial E_x}{\partial x} = -E_0 \frac{\alpha_1 \delta}{K_c^2} \left(\frac{m\pi}{a}\right) \sin\left(\frac{m\pi x}{a}\right) \sin\left(\frac{n\pi y}{b}\right) \sin\left(\frac{p\pi z}{d}\right)$$

$$= -E_0 \frac{\alpha_1 \delta}{K_c^2} \left(\frac{m\pi}{a}\right), \qquad (9.23)$$

$$\frac{\partial E_y}{\partial y} = -E_0 \frac{\alpha_2 \delta}{K_c^2} \left(\frac{n\pi}{b}\right) \sin\left(\frac{m\pi x}{a}\right) \sin\left(\frac{n\pi y}{b}\right) \sin\left(\frac{p\pi z}{d}\right)$$

$$= -E_0 \frac{\alpha_2 \delta}{K_c^2} \left(\frac{n\pi}{b}\right), \qquad (9.24)$$

which, compared with Eq. (9.20), yield

$$-\frac{\delta}{k_c^2}\left[\alpha_1\left(\frac{m\pi}{a}\right)+\alpha_2\left(\frac{n\pi}{b}\right)\right]=\left(\frac{p\pi}{d}\right),\qquad(9.25)$$

or

$$\delta=-\left(\frac{p\pi}{d}\right),\qquad(9.26)$$

and

$$k_c^2=\alpha_1\left(\frac{m\pi}{a}\right)+\alpha_2\left(\frac{n\pi}{b}\right)=\left(\frac{m\pi}{a}\right)^2+\left(\frac{n\pi}{b}\right)^2,\qquad(9.27)$$

having set

$$\alpha_1=\left(\frac{m\pi}{a}\right),\quad\text{and}\quad\alpha_2=\left(\frac{n\pi}{b}\right).\qquad(9.28)$$

Therefore Eqs. (9.21) and (9.22) become

$$E_x=-\frac{E_0}{k_c^2}\left(\frac{m\pi}{a}\right)\left(\frac{p\pi}{d}\right)\cos\left(\frac{m\pi x}{a}\right)\sin\left(\frac{n\pi y}{b}\right)\sin\left(\frac{p\pi z}{d}\right),\qquad(9.29)$$

and

$$E_y=-\frac{E_0}{k_c^2}\left(\frac{n\pi}{b}\right)\left(\frac{p\pi}{d}\right)\sin\left(\frac{m\pi x}{a}\right)\cos\left(\frac{n\pi y}{b}\right)\sin\left(\frac{p\pi z}{d}\right).\qquad(9.30)$$

The next step will be to find the magnetic field components using these electric field components. This task is accomplished by Ampere's law:

$$\vec{\nabla}\times\vec{H}=\vec{j}+\frac{\partial\vec{D}}{\partial t}\implies(\vec{j}=0)\implies\vec{\nabla}\times\vec{B}=\mu_0\varepsilon_0\frac{\partial\vec{E}}{\partial t}=\frac{1}{c^2}\frac{\partial\vec{E}}{\partial t}.\qquad(9.31)$$

Bearing in mind that B_z is null, as is typical for the TM mode, from Eq. (9.31) we can deduce that

$$\vec{\nabla}\times\vec{B}=\begin{pmatrix}\hat{x}&\hat{y}&\hat{z}\\\frac{\partial}{\partial x}&\frac{\partial}{\partial y}&\frac{\partial}{\partial z}\\B_x&B_y&0\end{pmatrix}=\begin{pmatrix}-\frac{\partial B_y}{\partial z}\\\frac{\partial B_x}{\partial z}\\\frac{\partial B_y}{\partial x}-\frac{\partial B_x}{\partial y}\end{pmatrix}=\frac{1}{c^2}\begin{pmatrix}\frac{\partial E_x}{\partial t}\\\frac{\partial E_y}{\partial t}\\\frac{\partial E_z}{\partial t}\end{pmatrix}.\qquad(9.32)$$

The time dependence of E_z has to be the same as the z dependence, i. e. of the cosine type. Hence we have

$$\frac{\partial B_y}{\partial z}=\frac{\omega}{c^2}\frac{E_0}{k_c^2}\left(\frac{m\pi}{a}\right)\left(\frac{p\pi}{d}\right)\cos\left(\frac{m\pi x}{a}\right)\sin\left(\frac{n\pi y}{b}\right)\sin\left(\frac{p\pi z}{d}\right)\sin(\omega t).\qquad(9.33)$$

Since we are mainly interested in the spatial structure of the standing waves, we may leave aside $\sin(\omega t)$, thus obtaining

$$\begin{aligned}B_y&=\frac{\omega}{c^2}\frac{E_0}{k_c^2}\left(\frac{m\pi}{a}\right)\cos\left(\frac{m\pi x}{a}\right)\sin\left(\frac{n\pi y}{b}\right)\left[\int\sin\left(\frac{p\pi z}{d}\right)d\left(\frac{p\pi z}{d}\right)\right],\\&=-\frac{\omega}{c^2}\frac{E_0}{k_c^2}\left(\frac{m\pi}{a}\right)\cos\left(\frac{m\pi x}{a}\right)\sin\left(\frac{n\pi y}{b}\right)\cos\left(\frac{p\pi z}{d}\right).\end{aligned}\qquad(9.34)$$

Similarly we have

$$B_x = \frac{\omega}{c^2} \frac{E_0}{k_c^2} \left(\frac{n\pi}{b}\right) \sin\left(\frac{m\pi x}{a}\right) \cos\left(\frac{n\pi y}{b}\right) \cos\left(\frac{p\pi z}{d}\right). \tag{9.35}$$

The Helmholtz equation (9.11) is deduced from the d'Alembert equation

$$\nabla^2 g(x, y, z, t) - \frac{1}{c^2} \frac{\partial^2}{\partial t^2} g(x, y, z, t) = 0, \tag{9.36}$$

for a time dependence of the oscillating type; in fact

$$g(x, y, z, t) = f(x, y, z) \exp(j\omega t) \quad \Longrightarrow \quad \frac{\partial^2 g}{\partial t^2} = f(j\omega)^2 e^{j\omega t} = -\omega^2 f e^{j\omega t}, \tag{9.37}$$

and Eq. (9.36) transforms into

$$\left[\nabla^2 f(x, y, z) + \left(\frac{\omega}{c}\right)^2\right] e^{j\omega t} = 0, \tag{9.38}$$

which coincides with Eq. (9.11) simply setting

$$k = \frac{\omega}{c} = \frac{2\pi f}{c} = \frac{2\pi}{\lambda}, \tag{9.39}$$

or

$$\omega = \frac{2\pi c}{\lambda} = c\sqrt{\left(\frac{m\pi}{a}\right)^2 + \left(\frac{n\pi}{b}\right)^2 + \left(\frac{p\pi}{d}\right)^2}, \tag{9.40}$$

having recalled Eq. (9.19). The lowest mode is M_{110} which reduces the quantities (9.27) and (9.40) to

$$k_c = \sqrt{\left(\frac{\pi}{a}\right)^2 + \left(\frac{\pi}{b}\right)^2}, \quad \text{and} \quad \omega = c\sqrt{\left(\frac{\pi}{a}\right)^2 + \left(\frac{\pi}{b}\right)^2} = cK_c. \tag{9.41}$$

Moreover, all the field components found above reduce to

$$E_x = E_y = 0, \tag{9.42}$$

$$E_z = E_0 \sin\left(\frac{\pi x}{a}\right) \sin\left(\frac{\pi y}{b}\right) = E_0 \sin\left(\frac{\pi x}{w}\right) \sin\left(\frac{\pi y}{w}\right), \tag{9.43}$$

$$B_x = \frac{E_0}{\omega} \frac{\pi}{b} \sin\left(\frac{\pi x}{a}\right) \cos\left(\frac{\pi y}{b}\right) = \frac{E_0}{\omega} \frac{\pi}{w} \sin\left(\frac{\pi x}{w}\right) \cos\left(\frac{\pi y}{w}\right), \tag{9.44}$$

$$B_y = \frac{E_0}{\omega} \frac{\pi}{a} \cos\left(\frac{\pi x}{a}\right) \sin\left(\frac{\pi y}{b}\right) = \frac{E_0}{\omega} \frac{\pi}{w} \cos\left(\frac{\pi x}{w}\right) \sin\left(\frac{\pi y}{w}\right), \tag{9.45}$$

$$B_y = 0, \tag{9.46}$$

having made the assessment $a = b = w$ (and $d = l$) which further yields

$$Q = \pi\sqrt{2}\frac{c}{w}. \tag{9.47}$$

Let us recall the definition of the quality factor:

$$Q = \frac{\omega U}{\langle P_{\text{loss}} \rangle}, \qquad \text{(CM: 9.87)}$$

where

$$U = \frac{1}{2}\varepsilon_0 \iiint (\vec{E} \cdot \vec{E}) d^3\vec{r} + \frac{1}{2\mu_0} \iiint (\vec{B} \cdot \vec{B}) d^3\vec{r} = U_{\text{E}} + U_{\text{M}} \qquad (9.48)$$

is the energy stored (CM: 9.81) in the cavity, and

$$\langle P_{\text{loss}} \rangle = \frac{R_s}{2\mu_0^2} \iint B_{\parallel}^2 dS \qquad (9.49)$$

is the time-averaged power loss (Problem 9–2) inside the conducting walls, where $\vec{B}_{\parallel}$ is the magnetic field close to the cavity walls. Bearing in mind Eqs. (9.43), (9.44), (9.45) from Eq. (9.48) we obtain respectively

$$U_{\text{E}} = \frac{1}{2}\varepsilon_0 \int_0^l dz \int_0^w \sin^2\left(\frac{\pi x}{w}\right) dx \int_0^w \sin^2\left(\frac{\pi y}{w}\right) dy = \frac{1}{8}\varepsilon_0 E_0 l w^2, \quad (9.50)$$

and

$$U_{\text{M}} = \frac{1}{2\mu_0}\left(\frac{E_0\pi}{\omega w}\right)^2 \int_0^l dz \left\{ \int_0^w \int_0^w \left[\sin^2\left(\frac{\pi x}{w}\right)\cos^2\left(\frac{\pi y}{w}\right) \right. \right.$$
$$\left. \left. + \cos^2\left(\frac{\pi x}{w}\right)\sin^2\left(\frac{\pi y}{w}\right) \right] dx\, dy \right\}. \quad (9.51)$$

Evaluating aside the expression within braces, we have

$$\{\dots\} = \int_0^w \left[\cos^2\left(\frac{\pi y}{w}\right) \int_0^w \sin^2\left(\frac{\pi x}{w}\right) dx \right.$$
$$\left. + \sin^2\left(\frac{\pi y}{w}\right) \int_0^w \cos^2\left(\frac{\pi x}{w}\right) dx \right] dy$$
$$= \frac{w}{2} \int_0^w \left[\cos^2\left(\frac{\pi y}{w}\right) + \sin^2\left(\frac{\pi y}{w}\right) \right] dy = \frac{w^2}{2}, \qquad (9.52)$$

or

$$U_{\text{M}} = \frac{l}{4\mu_0}\frac{E_0^2\pi^2}{\omega^2} = \frac{1}{16}\varepsilon_0 E_0^2 l \left(\frac{2\pi c}{\omega}\right)^2 = \frac{1}{16}\varepsilon_0 E_0^2 l \lambda^2. \qquad (9.53)$$

Inserting Eqs. (9.50) and (9.53) into Eq. (9.48), we finally obtain

$$U = \frac{1}{16}\varepsilon_0 E_0^2\, l \left(2w^2 + \lambda^2\right). \qquad (9.54)$$

In order to evaluate the power loss we must consider the magnetic field lines which gaze the cavity walls. With our mode we have the following situation:

- $B_x \hat{x}$ skims over the wall bounded by the x-z axes and over the parallel wall lying at a distance w;
- $B_y \hat{y}$ skims over the wall bounded by the y-z axes and over the parallel wall lying at a distance w;

This allows us to write

$$b_x(x,t) = B_x(y=0,w) = \frac{E_0}{w} \frac{\pi}{w} \sin\left(\frac{\pi x}{w}\right) \sin(\omega t), \qquad (9.55)$$

$$b_y(y,t) = B_y(x=0,w) = -\frac{E_0}{w} \frac{\pi}{w} \sin\left(\frac{\pi y}{w}\right) \sin(\omega t). \qquad (9.56)$$

Hence, reffering to Eq. (9.49), we have:

$$\frac{2\mu_0^2}{R_s} P_{\text{loss}}(\vec{r},t) = 2 \int_0^w b_x^2 dx \int_0^w dy + 2 \int_0^w dx \int_0^w b_y^2 dy = \frac{2l E_0^2 \pi^2}{\omega^2} \sin^2(\omega t),$$
$$(9.57)$$

and

$$\langle P_{\text{loss}} \rangle = \frac{1}{8} \varepsilon_0 E_0^2 \left(\frac{2\pi c}{\omega}\right)^2 \frac{R_s}{\mu_0} = \frac{1}{8} \varepsilon_0 E_0^2 \lambda^2 \frac{R_s}{\mu_0}. \qquad (9.58)$$

Recalling the definition (CM: 9.87) of the quality factor and taking into account Eqs. (9.54) and (9.58), we obtain

$$Q = \frac{\omega U}{\langle P_{\text{loss}} \rangle} = \frac{2w^2 + \lambda^2}{2\lambda^2} \frac{l\mu_0\omega}{R_s} \sim \frac{l\mu_0\omega}{R_s} = l\sqrt{2\sigma\mu_0\omega}, \qquad (9.59)$$

having considered Eq. (CM: 9.17)

$$R_s = \sqrt{\frac{\mu\omega}{2\sigma}} = \frac{1}{\sigma\omega}, \qquad \text{(CM: 9.17)}$$

and the definitions about it. It is worthwhile to emphasize how the larger the conductivity σ (or the smaller the resistivity) is, the larger the Q factor will be will be. This is strongly confirmed by superconducting cavities with their conductivities extremely small and their quality factors of the order of several billions.

9.4 Problem 9–4: Impedance of coupling loop

Derive Eq. (CM: 9.91)

$$Z_{\text{eff}} \approx i\omega L_1 - i\frac{\omega_0^2 M^2}{2L_2\left(\omega - \omega_0 - i\frac{\omega_0}{2Q}\right)}, \qquad \text{(CM: 9.91)}$$

for the effective impedance of a coupling loop connected to a simple cavity.

Solution:

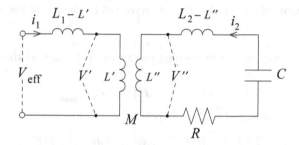

Fig. 9.3 Simple schematic of a transmission line coupled to a cavity using a loop of inductance L_1. The cavity is modeled by a lumped LRC circuit with respective values of L_2, R, and C for the inductance, resistance, and capacitance of the cavity. Part of the loop and cavity wall form a 1:1 turn transformer with mutual inductance $M = \sqrt{L'L''}$. The voltages across the inductances L' and L'' are respectively labeled V' and V''.

We would like to find $Z_{\text{eff}} = \hat{V}_{\text{eff}}/\hat{I}_1$ where the "hat" symbol indicates a Fourier transform of the variable. We may also write

$$V' = L'\frac{dI_1}{dt} + M\frac{dI_2}{dt}, \qquad (9.60)$$

$$V'' = M\frac{dI_1}{dt} + L''\frac{dI_2}{dt}. \qquad (9.61)$$

Multiplying the first equation by L'' and the second by M and then subtracting yields

$$L''V' - MV'' = (L'L'' - M^2)\frac{dI_1}{dt} = 0, \qquad (9.62)$$

so we find a formula equivalent to the turns ratio for a transformer:

$$\frac{V''}{V'} = \frac{L''}{M} = \sqrt{\frac{L''}{L'}}. \qquad (9.63)$$

Applying Kirchhoff's second law to the two loops in Fig. 9.3, we have

$$V_{\text{eff}} = L_1 \frac{dI_1}{dt} + M \frac{dI_2}{dt}, \tag{9.64}$$

$$0 = RI_2 + \frac{q_2}{C} + L_2 \frac{dI_2}{dt} + M \frac{dI_1}{dt}. \tag{9.65}$$

Taking the Fourier transforms of the this pair of equations yields

$$\hat{V}_{\text{eff}} = i\omega(L_1 \hat{I}_1 + M \hat{I}_2), \tag{9.66}$$

$$0 = \left(R + \frac{1}{i\omega C} + i\omega L_2 \right) \hat{I}_2 + i\omega M \hat{I}_1, \tag{9.67}$$

and combining to eliminate $\hat{I}_2$ gives

$$Z_{\text{eff}} = \frac{\hat{V}_{\text{eff}}}{\hat{I}_1} = i\omega L_1 + \frac{\omega^2 M^2}{R + \frac{1}{i\omega C} + i\omega L_2}. \tag{9.68}$$

For the uncoupled case ($M = 0$), the resonant frequency of the cavity is

$$\omega_0 = \frac{1}{L_2 C}. \tag{9.69}$$

The stored energy in the uncoupled cavity (with V_C across the capacitor) is

$$U = \frac{1}{2}(CV_C^2 + L_2 I_2^2) = \frac{1}{2} L_2 I_{2,\text{max}}^2, \tag{9.70}$$

and power loss averaged over one cycle is

$$\langle P_{\text{loss}} \rangle = \frac{\omega}{2\pi} \int_0^{2\pi/\omega} R\left[I_{2,\text{max}} \cos(\omega t) \right]^2 dt = \frac{1}{2} R I_{2,\text{max}}^2. \tag{9.71}$$

Recalling the definition of quality factor:

$$Q = \frac{U\omega}{\langle P_{\text{loss}} \rangle}, \tag{CM: 9.87}$$

we find

$$R = \frac{\omega L_2}{Q}. \tag{9.72}$$

We may now rewrite Eq. (9.68) as

$$Z_{\text{eff}} = \frac{\hat{V}_{\text{eff}}}{\hat{I}_1} = i\omega L_1 + \frac{i\omega^3 M^2 C}{i\omega C \frac{\omega L_2}{Q} + 1 - \omega^2 L_2 C}$$

$$= i\omega L_1 + \frac{i\omega^3 M^2 C}{i \frac{\omega^2}{\omega_0^2 Q} + \frac{\omega_0^2 - \omega^2}{\omega_0^2}}$$

$$= i\omega L_1 - i \frac{\omega^3 M^2}{L_2 \left((\omega - \omega_0)(\omega + \omega_0) - i \frac{\omega^2}{Q} \right)} \tag{9.73}$$

$$\approx i\omega L_1 - i \frac{\omega_0^2 M^2}{2 L_2 \left(\omega - \omega_0 - i \frac{\omega_0}{2Q} \right)}, \tag{9.74}$$

if we assume that ω is close to ω_0.

9.5   Problem 9–5: RFQ potential

Using Laplace's equation in cylindrical coordinates, show that Eq. (CM: 9.122)

$$\Phi(r,\theta,z) = \frac{V}{2}\left[\sum_{n=0}^{\infty} A_{0n} r^{2n} \cos 2n\theta\right.$$

$$\left. + \sum_{n=0}^{\infty}\sum_{l=1}^{\infty} A_{ln} r^{2n} I_{2n}(lkr) \cos 2n\theta \cos lkz\right], \quad \text{(CM: 9.122)}$$

is the general solution for the symmetry of an RFQ.

Solution:

Recall that the Laplace equation in cylindrical coordinates is

$$\nabla^2 \Phi = \frac{1}{r}\frac{\partial}{\partial r}\left(r\frac{\partial \Phi}{\partial r}\right) + \frac{1}{r^2}\frac{\partial^2 \Phi}{\partial \theta^2} + \frac{\partial^2 \Phi}{\partial z^2} = 0. \quad \text{(CM: 9.121)}$$

In cylindrical coordinates the Laplacian operator is separable, so we may look for solutions of the form:

$$\Phi_j \sim R(r)\,\Theta(\theta)\,Z(z). \quad (9.75)$$

So substituting into the Laplace equation and then dividing by Φ_j and multiplying by r^2, we find

$$\frac{1}{R}\left(r^2\frac{d^2R}{dr^2} + r\frac{dR}{dr}\right) + \frac{1}{\Theta}\frac{d^2\Theta}{d\theta^2} + \frac{r^2}{Z}\frac{d^2Z}{dz^2} = 0. \quad (9.76)$$

Note that in the empty region between the electrodes of the RFQ we want bounded functions. Clearly $\Theta(\theta)$ must be periodic, so that we may write

$$\frac{1}{\Theta}\frac{d^2\Theta_n}{d\theta^2} = -n^2, \quad (9.77)$$

and in general we expect something like

$$\Theta_n(z) = C_1 \cos n\theta + C_2 \sin n\theta. \quad (9.78)$$

Actually for quadrupole geometry with mirror symmetry in the horizontal-longitudinal (xz) and vertical-longitudinal (yz) planes, we can eliminate the sine functions and odd cosine functions so that the only functions functions are of the form

$$\Theta_{2n}(\theta) \sim \cos 2n\theta. \quad (9.79)$$

We should note that if there are misalignments or machining errors of the vanes, then unallowed (sine and odd $(2n + 1)$ cosine) harmonics could also be present. (See the similar discussion of allowed and unallowed harmonics for magnets in CM: § 4.1.)

Likewise for Z_l, we may look for periodic sine-like functions of z with

$$Z_l(z) \sim \cos(lkz + \phi_l). \tag{9.80}$$

We could have added a constant phase ϕ_{ln} to the lkz argument of the last cosine in Eq. (CM: 9.122) to allow for completeness of the general solution for allowed harmonics. Again we want bounded solutions rather than the unbounded hyperbolic functions. The differential equation from $Z_l(z)$ becomes

$$\frac{r^2}{Z} \frac{d^2 Z_l}{dz} = -l^2 k^2, \tag{9.81}$$

and combining this with Eqs. (9.79 and 9.76) leads to

$$r^2 \frac{d^2 R}{dr^2} + r \frac{dR}{dr} - [n^2 + l^2(kr)^2]R = 0, \tag{9.82}$$

or

$$\rho^2 \frac{d^2 R}{d\rho^2} + r \frac{dR}{d\rho} - [n^2 + \rho^2]R = 0, \tag{9.83}$$

with $\rho = lkr$. This last equation is the standard differential equation for the Modified Bessel Functions[7] $K_n(\rho)$ and $I_{\pm n}(\rho)$. Since the $K_n(\rho)$ are infinite at $\rho = 0$, we only need to consider the $I_n(\rho)$ for bounded solutions in the channel of the RFQ. So putting all this together we get the general solution for allowed harmonics as given by Eq. (CM: 9.122) allowing for the additional extra phase angles ϕ_{nl} with the $\cos(lkz)$ replaced by $\cos(lkz + \phi_{nl})$. In practice we would design the vanes for $\phi_{nl} = 0$, and reduce all the terms except the two given in Eq. (CM: 9.124):

$$\Phi = \frac{V}{2}[A_{01} r^2 \cos 2\theta + A_{10} I_0(kr) \cos kz]. \tag{CM: 9.124}$$

Of course, another way to "show" this would have been just to verify that Eq. (CM: 9.122) satisfies the Laplace equation.

9.6   Problem 9–6: RFQ phase oscillations

Derive an equation of motion for the longitudinal phase oscillations about the synchronous particle in an RFQ. For what values of ϕ_0 are small oscillations stable? What is the frequency of the small oscillations? (Hint: Construct difference equations for the deviations in phase and $W = -\Delta U/\omega_{\mathrm{rf}}$ for a test particle passing through one cell.)

Solution:

Recall (with a slight correction to add the missing subscript on ΔU) Eq. (CM: 9.133):

$$\Delta U_s = \int_0^{\beta\lambda_{\mathrm{rf}}/2} qE_z(0,0,z,t)\,dz$$

$$= \frac{qkA_{10}V}{2} I_0(0) \int_0^{\beta\lambda_{\mathrm{rf}}/2} \sin kz \sin(kz + \phi_s)\,dz$$

$$= \frac{\pi q A_{10}V}{4} \cos(\phi_s). \qquad\qquad \text{(CM: 9.133)}$$

For a particle with phase $\phi = \phi_s + \varphi$ this becomes

$$\Delta U = \frac{\pi q A_{10}V}{4} \cos(\phi_s + \varphi), \qquad\qquad (9.84)$$

and following the treatment given in CM: § 7.4, we may write

$$\Delta(\delta U) = \Delta U - \Delta U_s = \frac{\pi q A_{10}V}{4} (\cos\phi - \cos\phi_s), \qquad\qquad (9.85)$$

for the energy increment through one unit cell of the RFQ. The time of flight of the synchronous particle through the cell is

$$\tau = \frac{\pi}{\omega_{\mathrm{rf}}}, \qquad\qquad (9.86)$$

and also recalling the canonical momentum variable conjugate to the phase

$$W = -\delta U/\omega_{\mathrm{rf}} = -\frac{U - U_s}{\omega_{\mathrm{rf}}}, \qquad\qquad \text{(CM: 7.24)}$$

we construct the motion equation:

$$\dot{W} = \frac{dW}{dt} \simeq \frac{\Delta(\delta U)}{\tau} = \frac{q A_{10}}{4} (\cos\phi_s - \cos\phi). \qquad\qquad (9.87)$$

Since the designed path of the synchronous particle is a straight line, the momentum compaction for the RFQ is zero and we have

$$-\frac{d\tau}{\tau} = \gamma^{-2} \frac{dp}{p}. \qquad\qquad (9.88)$$

The difference equation for the phase is then

$$\Delta\varphi \simeq \frac{d\varphi}{dt}\,\tau = \omega_{\rm rf}\,\delta t \qquad \text{(CM: 7.26)}$$

$$= -\frac{\omega_{\rm rf}}{\gamma^2}\frac{dp}{p}\,\tau, \qquad (9.89)$$

but

$$\frac{dp}{p} = \frac{dU}{\beta^2 U_s} = -\frac{\omega_{\rm rf}W}{\beta^2 U_s}. \qquad (9.90)$$

Combining Eqs. (9.88–9.90) to evaluate $\dot\varphi \simeq \Delta\varphi/\tau$, the phase derivative is

$$\frac{d\varphi}{dt} = \frac{\omega_{\rm rf}{}^2}{\beta^2\gamma^2 U_s}\,W. \qquad \text{(CM: 9.104)}$$

Combining Eqs. (9.87 and CM: 9.104) yields the motion equation:

$$\ddot\phi + \frac{qA_{10}V}{4}\frac{\omega_{\rm rf}{}^2}{\gamma^2\beta^2 U_s}\big(\cos\phi - \cos\phi_s\big) = 0. \qquad (9.91)$$

This has a very similar form to Eq. (CM: 7.30):

$$\ddot\varphi + \frac{h\omega_s^2\eta_{\rm tr}qV}{2\pi\beta^2 U_s}(\sin\phi - \sin\phi_s) = 0, \qquad \text{(CM: 7.30)}$$

except that we have cosine functions rather than sine functions, although this has been explained in CM: § 9.5. Since $\cos\phi = \sin(\phi + \pi/2)$, the stability requirements will be identical to those of CM: § 7.5 provided that we shift the synchronous phase by 90°:

$$\ddot\varphi + \frac{\Omega_s^2}{\cos\phi_s}[\sin(\varphi + \phi_s) - \sin\phi_s] = 0, \qquad \text{(CM: 7.33)}$$

$$\phi_s \to \phi_s + \frac{\pi}{2}, \qquad (9.92)$$

$$\ddot\varphi + \frac{\Omega_s^2}{-\sin\phi_s}[\cos(\varphi + \phi_s) - \cos\phi_s] = 0, \qquad (9.93)$$

however we should remember that the Ω_s^2 of Eq. (CM: 7.33) contains a factor of $\cos\phi_s \to -\sin\phi_s$.

For small oscillations ($|\varphi| \ll 0$):

$$\cos(\phi_s + \varphi) - \cos\phi_s = \cos\phi_s\cos\varphi - \sin\phi_s\sin\varphi - \cos\phi_s \simeq -\sin\phi_s\,\varphi, \qquad (9.94)$$

and we obtain

$$\ddot\varphi + \Omega_s^2\varphi \simeq 0, \qquad (9.95)$$

with

$$\Omega_s = \sqrt{\frac{\omega_{\rm rf}{}^2 qA_{10}V(-\sin\phi_s)}{4\gamma^2\beta^2 U_s}}, \qquad (9.96)$$

so that for stability we must have $qA_{10}V\sin\phi_s < 0$.

Chapter 10

Problems of Chapter 10: Resonances

Before proceeding to the problems, we would like to make a few general comments on the particular solutions of driven oscillation equations. Hopefully these are fairly obvious, but if not, the recap of the methods of CM: § 10.1 may help.

Consider two similar equations with different driving terms $f_1(\theta)$, and $f_2(\theta)$ but identical homogeneous parts:

$$\frac{d^2x}{d\theta^2} + Q^2x = f_1(\theta), \tag{10.1}$$

$$\frac{d^2x}{d\theta^2} + Q^2x = f_2(\theta), \tag{10.2}$$

with the respective inhomogeneous solutions x_{p1} and x_{p2}. We may easily add the two equations having substituted their respective particular solutions to obtain

$$\frac{d^2}{d\theta^2}(x_{p1} + x_{p2}) + Q_{\mathrm{H}}{}^2(x_{p1} + x_{p2}) = f_1(\theta) + f_2(\theta), \tag{10.3}$$

Of course this is easily extended to any sum of driving terms so that the particular solution for the equation with summed driving terms is just the sum of the particular solutions for the individual driving terms:

$$\frac{d^2x_p}{dx^2} + Q^2x_p = \sum_{j=1}^{N} f_j(\theta), \quad \text{with} \quad x_p = \sum_{j=1}^{N} x_{pj}, \tag{10.4}$$

where each x_{pj} is a particular solution to its respective equation:

$$\frac{d^2x_{pj}}{dx^2} + Q^2x_{pj} = f_j(\theta). \tag{10.5}$$

Assuming that a general driving function can be expanded in a Fourier series

$$f(\theta) = \sum_{n=-\infty}^{\infty} A_j e^{in\theta}, \tag{10.6}$$

then we may sum the particular solutions for each Fourier harmonic to obtain the overall particular solution for the general driving function $f(\theta)$. So for the nth harmonic $e^{in\theta}$ we have a particular solution:

$$x_{pn} = \frac{A_n(e^{in\theta})}{Q^2 - n^2},$$ (10.7)

which is easily be verified by substitution:

$$\left(\frac{d^2}{dx^2} + Q^2\right)\frac{A_n e^{in\theta}}{Q^2 - n^2} = (-n^2 + Q^2)\frac{A_n e^{in\theta}}{Q^2 - n^2} = A_n e^{in\theta}.$$ (10.8)

More generally, we should also note that

$$x_{p\alpha} = \frac{A(e^{i\alpha\theta})}{Q^2 - \alpha^2},$$ (10.9)

will be a particular solution of the inhomogeneous equation

$$\frac{d^2x}{d\theta^2} + Q^2 x = A e^{i\alpha\theta},$$ (10.10)

for any real number α. We may now add in a bit of the homogeneous solution $-e^{iQ\theta}$ to form another inhomogeneous solution

$$
\begin{aligned}
x_{p\alpha a} &= \frac{A(e^{i\alpha\theta} - e^{iQ\theta})}{Q^2 - \alpha^2} \\
&= \frac{A}{Q^2 - \alpha^2} e^{i\frac{Q+\alpha}{2}\theta}\left(e^{-i\frac{Q-\alpha}{2}\theta} - e^{i\frac{Q-\alpha}{2}\theta}\right) \\
&= \frac{-2iA\,e^{i\frac{Q+\alpha}{2}\theta}}{Q^2 - \alpha^2}\sin\left(\frac{Q-\alpha}{2}\theta\right)
\end{aligned}
$$ (10.11)

Taking the real part of $x_{p\alpha a}$ yields a solution to the equation

$$\frac{d^2x}{d\theta^2} + Q^2 x = A\cos(\alpha\theta),$$ (10.12)

namely

$$x_{par} = \text{Re}(x_{p\alpha a}) = \frac{A}{2}\frac{\sin\left(\frac{Q+\alpha}{2}\theta\right)}{\frac{Q+\alpha}{2}}\frac{\sin\left(\frac{Q-\alpha}{2}\theta\right)}{\frac{Q-\alpha}{2}},$$ (10.13)

which demonstrates linear growth in amplitude for resonances at $Q = \pm\alpha$.

10.1 Problem 10–1: Skew multipole driven resonances

Find the resonance relations for the skew multipoles and make plots of the resonance relations for the quadrupole, sextupole, octopole, and decapoles.

Solution:

Recall Eq. (CM: 4.8):

$$B_x - iB_y = B_0 \sum_{n=0}^{\infty} (a_n - ib_n) \left(\frac{x + iy}{a} \right)^n. \tag{CM: 4.8}$$

For simplicity let us absorb the B_0 and a^{-n} factors into the multipole coefficients to quell the algebraic clutter, so this becomes:

$$B_x - iB_y = \sum_{n=0}^{\infty} (a_n - ib_n)(x + iy)^n. \tag{10.14}$$

For the skew quadrupole (a_1 only), this gives

$$B_x = a_1 x, \quad \text{and} \quad B_y = -a_1 y \tag{10.15}$$

Our horizontal motion is driven by the vertical field, so we will should examine the equations

$$\frac{d^2 x}{d\theta^2} + Q_H{}^2 x = \epsilon \cos(m\theta) \left(\frac{\partial B_y}{\partial y} \right) y = -\epsilon a_1 \cos(m\theta) y, \tag{10.16}$$

$$\frac{d^2 y}{d\theta^2} + Q_V{}^2 y = \epsilon \cos(m\theta) \left(\frac{\partial B_x}{\partial x} \right) x = \epsilon a_1 \cos(m\theta) x, \tag{10.17}$$

This pair will give the linear coupling resonances discussed in CM: § 10.2. More generally, we could consider the system of equations with coupling:

$$\frac{d^2}{d\theta^2} \begin{pmatrix} x \\ y \end{pmatrix} + \begin{pmatrix} Q_H{}^2 x \\ Q_V{}^2 y \end{pmatrix} = \epsilon \cos(m\theta) \left[x \frac{\partial}{\partial x} + y \frac{\partial}{\partial y} \right] \begin{pmatrix} B_y \\ -B_x \end{pmatrix}, \tag{10.18}$$

where the minus before the B_x comes from evaluating $q\vec{v} \times \vec{B}$.

If we only consider the a_n multipole, and then take the partial derivatives Eq. (10.14) with respect to x and y we have

$$\frac{\partial B_x}{\partial x} - i \frac{\partial B_y}{\partial x} = n a_n (x + iy)^{n-1}, \tag{10.19}$$

$$\frac{\partial B_x}{\partial y} - i \frac{\partial B_y}{\partial y} = i n a_n (x + iy)^{n-1}, \tag{10.20}$$

so the individual derivatives must be

$$\frac{\partial B_x}{\partial y} = \frac{\partial B_y}{\partial x} = -na_n \text{Im}[(x+iy)^{n-1}], \tag{10.21}$$

$$\frac{\partial B_x}{\partial x} = -\frac{\partial B_y}{\partial y} = na_n \text{Re}[(x+iy)^{n-1}], \tag{10.22}$$

which satisfy both $\nabla \cdot \vec{B} = 0$ and $\nabla \times \vec{B} = 0$.

With only the a_n term, having set all others to zero, we have

$$x\frac{\partial B_y}{\partial x} + y\frac{\partial B_y}{\partial y} = -na_n \left\{ x\,\text{Im}\left[(x+iy)^{n-1}\right] + y\,\text{Re}\left[(x+iy)^{n-1}\right] \right\},$$

$$\tag{10.23}$$

for the horizontal (upper) part of Eq. (10.18), and

$$x\frac{\partial B_x}{\partial x} + y\frac{\partial B_x}{\partial y} = na_n \left\{ x\,\text{Re}\left[(x+iy)^{n-1}\right] - y\,\text{Im}\left[(x+iy)^{n-1}\right] \right\},$$

$$\tag{10.24}$$

for the vertical (lower) part.

The binomial theorem gives

$$(x+iy)^{n-1} = \sum_{j=0}^{n-1} \binom{n-1}{j} x^{n-1-j} i^j y^j, \tag{10.25}$$

which splits into the real and imaginary parts:

$$\text{Re}[(x+iy)^{n-1}] = \sum_{j=0}^{\left[\frac{n-1}{2}\right]} \left[(-1)^j \binom{n-1}{2j} x^{n-1-2j} y^{2j} \right], \tag{10.26}$$

$$\text{Im}[(x+iy)^{n-1}] = \sum_{j=1}^{\left[\frac{n}{2}\right]} \left[(-1)^{j-1} \binom{n-1}{2j-1} x^{n-2j} y^{2j-1} \right]$$

$$= \sum_{j=0}^{\left[\frac{n}{2}-1\right]} \left[(-1)^j \binom{n-1}{2j+1} x^{n-2j-2} y^{2j+1} \right]. \tag{10.27}$$

Hence Eq. (10.23) becomes

$$x\frac{\partial B_y}{\partial x} + y\frac{\partial B_y}{\partial y} = -na_n \left\{ \sum_{j=0}^{\left[\frac{n-2}{2}\right]} \left[(-1)^j \binom{n-1}{2j+1} x^{n-2j-1} y^{2j+1} \right] \right.$$

$$\left. + \sum_{j=0}^{\left[\frac{n-1}{2}\right]} \left[(-1)^j \binom{n-1}{2j} x^{n-2j-1} y^{2j+1} \right] \right\}, \tag{10.28}$$

with odd powers of y, while Eq. (10.24) transforms into

$$x \frac{\partial B_x}{\partial x} + y \frac{\partial B_x}{\partial y} = na_n \left\{ \sum_{j=0}^{\left[\frac{n-1}{2}\right]} (-1)^j \binom{n-1}{2j} x^{n-2j} y^{2j} \right.$$

$$\left. - \sum_{j=0}^{\left[\frac{n-2}{2}\right]} (-1)^j \binom{n-1}{2j+1} x^{n-2j-2} y^{2j+2} \right\}$$

$$= na_n \left\{ \sum_{j=0}^{\left[\frac{n-1}{2}\right]} (-1)^j \binom{n-1}{2j} x^{n-2j} y^{2j} \right.$$

$$\left. + \sum_{j=1}^{\left[\frac{n}{2}\right]} (-1)^j \binom{n-1}{2j-1} x^{n-2j} y^{2j} \right\}, \qquad (10.29)$$

with even powers of y. If we now make the substitutions

$$x = A_1 \cos(Q_H \theta), \quad \text{and} \quad y = A_2 \cos(Q_V \theta), \qquad \text{(CM:10.33)}$$

the horizontal equation has right-hand terms proportional to

$$\cos(m\theta) \cos^{n-(2j+1)}(Q_H \theta) \cos^{2j+1}(Q_V \theta), \quad \text{for } j : 0 \to \left[\frac{n-1}{2}\right], \qquad (10.30)$$

and the vertical equation has terms on the right like

$$\cos(m\theta) \cos^{n-(2j+2)}(Q_H \theta) \cos^{2j+2}(Q_V \theta), \quad \text{for } j : 0 \to \left[\frac{n-1}{2}\right]. \qquad (10.31)$$

Recall Eq. (CM: 10.36) for the series expansion of the product of the cosines:

$$\cos(m\theta) \cos^p(Q_H \theta) \cos^q(Q_V \theta)$$

$$= 2^{-(p+q)} \sum_{k=0}^{p} \sum_{l=0}^{q} \binom{p}{k} \binom{q}{l}$$

$$\times \cos \left\{ [(p - 2k)Q_H + (q - 2l)Q_V - m] \theta \right\}. \qquad \text{(CM:10.36)}$$

Eq. (10.30) with $p = n - (2j + 1)$, $q = 2j + 1$ gives resonance conditions

$$[n - 1 - 2(j + k) \pm 1]Q_H + [2(j - l) + 1]Q_V = m. \qquad (10.32)$$

Similarly Eq. (10.31) with $p = n - 2j$, $q = 2j$ yields resonances when

$$[(n - 2(j + k)]Q_H + [2(j - l) \pm 1]Q_V = m. \qquad (10.33)$$

Rather than get caught up in evaluating all the limits from the sums, it is perhaps simpler to evaluate Eqs. (10.23 and 10.24) for each value of n.

Table 10.1   Skew multipole polynomials

n	multipole	Eq. (10.23)	Eq. (10.24)
1	a_1	$-y$	x
2	a_2	$-2xy$	$x^2 - y^2$
3	a_3	$-3x^2y + y^3$	$x^3 - 3xy^2$
4	a_4	$4xy^3 - 4x^3y$	$x^4 - 6x^2y^2 + y^4$

Table 10.1 lists the polynomial expressions obtained from Eqs. (10.23 and 10.24) for skew quadrupole, sextupole, octopole, and decapole multipoles.

So we see that the skew quadrupole drives the resonances

$$\pm Q_{\mathrm{H}} \pm Q_{\mathrm{V}} = m, \tag{10.34}$$

i. e. the linear coupling resonances (sum and difference) of CM: §10.2.

For the skew sextupole:

$$\pm Q_{\mathrm{V}} = m, \tag{10.35}$$

$$\pm 2Q_{\mathrm{H}} \pm Q_{\mathrm{V}} = m, \tag{10.36}$$

$$\pm 3Q_{\mathrm{V}} = m. \tag{10.37}$$

For the skew octopole:

$$\pm Q_{\mathrm{H}} \pm Q_{\mathrm{V}} = m,$$

$$\pm 3Q_{\mathrm{H}} \pm Q_{\mathrm{V}} = m,$$

$$\pm Q_{\mathrm{H}} \pm 3Q_{\mathrm{V}} = m. \tag{10.38}$$

For the skew decapole:

$$\pm Q_{\mathrm{V}} = m,$$

$$\pm 2Q_{\mathrm{H}} \pm Q_{\mathrm{V}} = m,$$

$$\pm 3Q_{\mathrm{V}} = m,$$

$$\pm 4Q_{\mathrm{H}} \pm Q_{\mathrm{V}} = m,$$

$$\pm 2Q_{\mathrm{H}} \pm 3Q_{\mathrm{V}} = m,$$

$$\pm 5Q_{\mathrm{V}} = m. \tag{10.39}$$

Fig. 10.1 displays the resonance lines for the skew quadrupole, sextuple, octopole, and decapole multipoles.

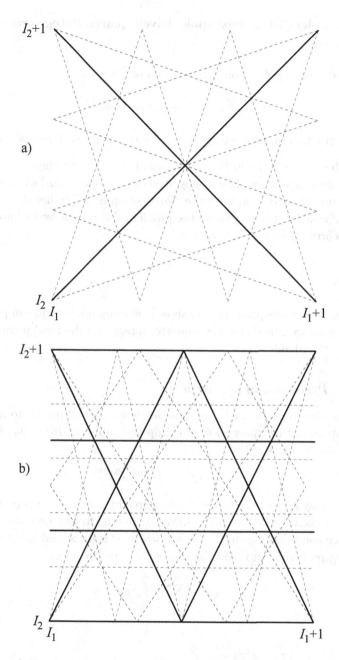

Fig. 10.1 a) Skew quad lines (solid) and skew octopole lines (bold and dashed). b) Skew sextupole (bold) and skew decapole (bold and dashed) lines.

10.2 Problem 10–2: Sextupole driven quarter-integer resonance

Show that a sextupole resonance equation of the type

$$\frac{d^2x}{d\theta^2} + Q_H{}^2 x = \varepsilon x^2 \cos(m\theta), \tag{10.40}$$

will give rise to a quarter-integer resonance condition for terms of order ε^2.

Note: The obvious typo in Eq. (10.40) has been corrected here.
Also, we must apologize. Due sloppy algebra on a blackboard when we first posed concocted problem, we wrote "terms of order ε^2" rather than ε^6. We probably would not have written the problem this way, if we had done our algebra correctly in the first place.

Solution:

The intention of this problem is to show how a sextupole term can produce a four resonance islands having a quarter-integer for the fixed point at the center of the islands.

10.2.1 *Perturbation approach*

For a simpler notation, let us use complex notation and primes to indicate a derivatives with respect to θ and drop the subscript from Q_H, so that Eq. (10.40) becomes

$$x'' + Q^2 x = \varepsilon e^{im\theta} x^2. \tag{10.41}$$

where we may extract the cosine driving force by taking the real part of the equation since $\text{Re}(e^{im\theta}) = \cos(m\theta)$. We are hoping to find a quarter-integer resonance condition like $4Q \pm Pm = 0$ where P is some odd integer.

Using perturbation theory, expand $x(\theta)$ in powers of ε:

$$x(\theta) = \sum_{n=0}^{\infty} \varepsilon^n x_n(\theta), \tag{10.42}$$

so

$$\begin{aligned}
x^2 &= (x_0 + \varepsilon x_1 + \varepsilon^2 x_2 + \cdots)^2 \\
&= x_0^2 + 2\varepsilon x_0 x_1 + \varepsilon^2(x_1^2 + 2x_0 x_2) + \varepsilon^3 2(x_0 x_3 + x_1 x_2) + \cdots. \tag{10.43}
\end{aligned}$$

Substituting into Eq. (10.41) we obtain the series of differential equations:

$$x_0'' + Q^2 x_0 = 0, \tag{10.44}$$

$$\varepsilon(x_1'' + Q^2 x_1) = \varepsilon e^{im\theta} \, x_0^2, = \varepsilon F_1(\theta) \tag{10.45}$$

$$\varepsilon^2(x_2'' + Q^2 x_2) = \varepsilon^2 e^{im\theta} \, 2x_0 x_1, = \varepsilon^2 F_2(\theta) \tag{10.46}$$

$$\varepsilon^3(x_3'' + Q^2 x_3) = \varepsilon^3 e^{im\theta} \, (x_1^2 + 2x_0 x_2) = \varepsilon^3 F_3(\theta), \tag{10.47}$$

$$\varepsilon^4(x_4'' + Q^2 x_4) = \varepsilon^4 e^{im\theta} \, 2(x_0 x_3 + x_1 x_2) = \varepsilon^4 F_4(\theta), \tag{10.48}$$

$$\vdots$$

Dividing out the powers of ε from each equation, we see that Eq. (10.44) is just a simple harmonic oscillator with a solution

$$x_0 = ae^{iQ\theta} + be^{-iQ\theta}. \tag{10.49}$$

For simplicity, let us just use the $e^{iQ\theta}$ part, taking $b = 0$, and assuming a is a real constant, since we do not need to examine all possible solutions in order to find the desired effect.

Eq. (10.45) is then a forced harmonic oscillator with a driving function of the form

$$F_1(\theta) = e^{im\theta}(ae^{iQ\theta})^2 = a^2 e^{i(2Q+m)\theta}, \tag{10.50}$$

with the frequency $|2Q + m|\omega_s$ where ω_s is 2π times the revolution frequency. If we had kept the negative exponent terms in Eq. (10.49), we would also find another two frequencies: $|2Q - m|\omega_s$, and $|m|\omega_s$. So now solving Eq. (10.45) we find from Eq. (10.7) with $n = 2Q + m$ the particular solution

$$x_{p1} = \frac{a^2 e^{i(2Q+m)\theta}}{Q^2 - (2Q + m)^2}. \tag{10.51}$$

Using a similar trick to that of CM: §1, we may add a homogeneous part to construct another particular solution

$$
\begin{aligned}
x_{p1a} &= \frac{a^2 e^{i(2Q+m)\theta} - e^{iQ\theta}}{Q^2 - (2Q + m)^2} \\
&= \frac{a^2 e^{i\frac{(2Q+m)+Q}{2}\theta} \left(e^{i\frac{(2Q+m)-Q}{2}\theta} - e^{i\frac{Q-(2Q+m)}{2}\theta} \right)}{[Q + (2Q + m)][Q - (2Q + m)]} \\
&= -2a^2 i \, \frac{e^{i\frac{3Q+m}{2}\theta}}{3Q + m} \frac{\sin\left(\frac{Q+m}{2}\theta \right)}{Q + m}
\end{aligned}
\tag{10.52}
$$

Taking the real part of this solution yields

$$x_{p1r} = \text{Re}(x_{p1a}) = \frac{a^2}{2} \frac{\sin\left(\frac{3Q+m}{2}\theta\right)}{\frac{3Q+m}{2}} \frac{\sin\left(\frac{Q+m}{2}\theta\right)}{\frac{Q+m}{2}}, \qquad (10.53)$$

which shows linear resonant growth for either $Q \to -m$ or $Q \to -m/3$.

Using $x_0 = ae^{iQ\theta}$ and $x_1 = x_{p1}$, now let us take in Eq. (10.46) the driving function

$$F_2(\theta) = e^{im\theta} 2x_0 x_{p1} = \frac{-2a^3 e^{i(3Q+2m)\theta}}{(Q+m)(3Q+m)}. \qquad (10.54)$$

Eq. (10.7) with $n = 3Q + 2m$ now yields the particular solution for Eq. (10.46):

$$\begin{aligned}
x_{p2} &= \frac{-2a^3}{(Q+m)(3Q+m)} \frac{e^{i(3Q+2m)\theta}}{Q^2 - (3Q+2m)^2} \\
&= \frac{2a^3}{(Q+m)(3Q+m)} \frac{e^{i(3Q+2m)\theta}}{(4Q+2m)(2Q+2m)}.
\end{aligned} \qquad (10.55)$$

With an additional homogeneous component subtracted as before, we find

$$x_{p2r} = \text{Re}\left(\frac{2a^3}{(Q+m)(3Q+m)} \frac{e^{i(3Q+2m)\theta} - e^{iQ\theta}}{(4Q+2m)(2Q+2m)}\right). \qquad (10.56)$$

$$e^{i(3Q+2m)\theta} - e^{iQ\theta} = e^{i\frac{(4Q+2m)}{2}\theta}\left(e^{-i\frac{(2Q+2m)}{2}\theta} - e^{-i\frac{2Q+2m}{2}\theta}\right), \qquad (10.57)$$

which adds resonant conditions for $2Q + m = 0$ and $Q + m = 0$.

D'oh![30] What can we say, but "oops".

To order ε^2 we have not found our desired "$4Q \pm Pm = 0$" type resonance.

This brings to mind something which Professor D. G. Ravenhall used to say in his quantum mechanics class at the University of Illinois: "If you can't solve the problem, change it until you can."

So we should now ask: Which order of ε^n produces a quarter-integer resonance?

Let us proceed to the 3rd order in ε with

$$F_3(\theta) = e^{im\theta}(x_{p1}^2 + 2x_0 x_{p2}). \qquad (10.58)$$

$$x_{p1}^2 = \frac{a^4 e^{i(4Q+2m)\theta}}{(3Q+m)^2(Q+m)^2}.$$

$$2x_0 x_{p2} = \frac{a^4 e^{i(4Q+2m)\theta}}{(Q+m)^2(2Q+m)(3Q+m)}. \qquad (10.59)$$

So we have

$$F_3(\theta) = \kappa_3 a^4 \, e^{i(4Q+3m)\theta}. \tag{10.60}$$

with

$$\kappa_3 = \frac{(5Q + 2m)}{(Q + m)^2(2Q + m)(3Q + m)^2} \tag{10.61}$$

which will be nonzero and finite so long as

$$Q \neq -\frac{2m}{3}, \quad -m, \quad -\frac{m}{2}, \quad \text{or} \quad -\frac{m}{3}, \tag{10.62}$$

none of which are quarter-integers. So we find a particular solution for the third order:

$$x_{p3} = \kappa_3 a^4 \frac{e^{i(4Q+3m)\theta}}{Q^2 - (4Q + 3m)^2}, \tag{10.63}$$

and with $\alpha = 4Q+3m$ in Eq. (10.13) which gives us a fifth-integer resonance $Q = -3m/5$.

To summarize things so far, for the equation

$$x''_{pn} + Q^2 x_{pn} = F_n(\theta), \tag{10.64}$$

with

$$F_n(\theta) = a^{n+1} \kappa_n e^{i\alpha_n \theta}, \tag{10.65}$$

$$\alpha_n = (n + 1)Q + nm, \tag{10.66}$$

$$\kappa_0 = 1, \tag{10.67}$$

$$\kappa_1 = 1, \tag{10.68}$$

$$\kappa_2 = \frac{-2}{(Q + m)(3Q + m)}, \tag{10.69}$$

$$\kappa_3 = \frac{5Q + 2m}{(Q + m)^2(2Q + m)(3Q + m^3)}, \tag{10.70}$$

and the constants κ_n to be found later, we have the solution

$$x_{pn} = a^{n+1} \kappa_n \frac{e^{i\alpha_n \theta}}{Q^2 - \alpha_n^2}. \tag{10.71}$$

Another particular solution with an added bit of homogeneous solution:

$$x_{p\alpha_n r} = \frac{a^{n+1} \kappa_n}{2} \frac{\sin\left(\frac{Q+\alpha_n}{2}\theta\right)}{\frac{Q+\alpha_n}{2}} \frac{\sin\left(\frac{Q-\alpha_n}{2}\theta\right)}{\frac{Q-\alpha_n}{2}}, \tag{10.72}$$

shows resonances for

$$0 = Q + \alpha_n = (n+2)Q + nm, \qquad (10.73)$$

$$0 = Q - \alpha_n = -n(Q+m). \qquad (10.74)$$

So from Eq. (10.73) we may expect an ε^n resonance when

$$Q = -\frac{n}{n+2}m, \qquad (10.75)$$

and we expect that $n = 6$ will give the lowest order quarter-integer resonance for $Q = -3m/4$. What remains is to verify that $\kappa_6 \neq 0$ when $Q = -3m/4$.

Listing the F_n functions with the polynomials in x_{pn} up to order 6, we have

$$F_1(\theta) = e^{im\theta}x_0^2, \qquad (10.76)$$

$$F_2(\theta) = e^{im\theta}2(x_0 x_1), \qquad (10.77)$$

$$F_3(\theta) = e^{im\theta}(x_1^2 + 2x_0 x_2), \qquad (10.78)$$

$$F_4(\theta) = e^{im\theta}2(x_0 x_3 + x_1 x_2), \qquad (10.79)$$

$$F_5(\theta) = e^{im\theta}(x_2^2 + 2x_0 x_4 + 2x_1 x_3), \qquad (10.80)$$

$$F_6(\theta) = e^{im\theta}2(x_0 x_5 + x_1 x_4 + x_2 x_3), \qquad (10.81)$$

which may be evaluated with a lot of recursive algebra via

$$\kappa_n = \frac{F_n(0)}{a^n}, \qquad (10.82)$$

A Maxima[2] script, listed in §10.2.2, gives the results:

$$\kappa_4 = -\frac{25Q + 13m}{3(Q+m)^3(2Q+m)(3Q+m)^2(5Q+3m)}, \qquad (10.83)$$

$$\kappa_5 = \frac{35Q^2 + 32Qm + 7m^2}{4(Q+m)^4(2Q+m)^2(3Q+m)^3(3Q+2m)}, \qquad (10.84)$$

$$\kappa_6 = -\frac{(35Q + 23m)(140Q^2 + 147Qm + 37m^2)}{60(Q+m)^5(2Q+m)^2(3Q+m)^3(3Q+2m)(5Q+3m)(7Q+5m)}. \qquad (10.85)$$

Of course what we needed to verify was that κ_6 is nonzero for $Q = -3m/4$:

$$\kappa_6|_{Q=-3m/4} = \frac{74973184}{5625m^{10}}. \qquad (10.86)$$

10.2.2   Maxima *code to evaluate* κ_n *up to* $n = 6$

```
/* Maxima script to evaluate the kappa_6 factor for Q=-3m/4

   For simplicity we have shortened some names,
   e.g.  x_{p1}=x1, etc.
   Also we have set a=1.

   Note also:
     If you paste this into an interface such as
     "wxMaxima", you may have to enter f5(t,Q) and f6(t,Q)
     below as a single line without the backslash
     continuation.

     "wxMaxima" also does not like comment lines like this.

     This script was run using the Linux (Ubuntu 10.4) OS.
*/

alpha(n,Q):=(n+1)*Q+n*m;

x0(t,Q):=exp(%i*Q*t);
x1(t,Q):=exp(%i*alpha(1,Q)*t)/(Q**2-alpha(1,Q)**2);

f2(t,Q):=exp(%i*m*t)*2*x0(t,Q)*x1(t,Q);
k2(Q):=f2(0,Q);
x2(t,Q):=k2(Q)*exp(%i*alpha(2,Q)*t)/(Q**2-alpha(2,Q)**2);

f3(t,Q):=exp(%i*m*t)*( x1(t,Q)**2 + 2*x0(t,Q)*x2(t,Q) );
k3(Q):=f3(0,Q);
x3(t,Q):=k3(Q)*exp(%i*alpha(3,Q)*t)/(Q**2-alpha(3,Q)**2);

f4(t,Q):=exp(%i*m*t)*2*( x0(t,Q)*x3(t,Q)+x1(t,Q)*x2(t,Q) );
k4(Q):=f4(0,Q);
x4(t,Q):=k4(Q)*exp(%i*alpha(4,Q)*t)/(Q**2-alpha(4,Q)**2);

f5(t,Q):=exp(%i*m*t) \
    * (x2(t,Q)**2+2*x0(t,Q)*x4(t,Q)+2*x1(t,Q)*x3(t,Q));
k5(Q):=f5(0,Q);
```

```
x5(t,Q):=k5(Q)*exp(%i*alpha(5,Q)*t)/(Q**2-alpha(5,Q)**2);

f6(t,Q):=exp(%i*m*t)*2*(x0(t,Q)*x5(t,Q)+x1(t,Q)*x4(t,Q) \
    +x2(t,Q)*x3(t,Q));
k6(Q):=f6(0,Q);
x6(t,Q):=k6(Q)*exp(%i*alpha(6,Q)*t)/(Q**2-alpha(6,Q)**2);


/* This is what we want: */

k6(-3*m/4);

factor(k3(Q));
factor(k4(Q));
factor(k5(Q));
factor(k6(Q));
```

At NAL (what Fermilab was called prior to 1974) evidence was seen that quarter-integer resonances could be driven by the sextupoles [17, 53, 58]. Later, a beam experiment at the Tevatron demonstrated that the beam could be kicked into resonance islands [14, 60], while at the IUCF storage at Indiana University, the $4Q_V = 15$ was studied [41, 40, 68].

10.2.3 *Simulation approach*

In this section, we present a few examples from a numerical simulation program written in the C language[37] and listed in §10.2.4. The program does a simple symplectic integration by breaking each turn into N_{step} steps with $\mu_{\text{step}} = 2\pi Q_0/N_{\text{step}}$ and applying a nonlinear thin lens kick at the beginning of each step:

$$\begin{pmatrix} x_{j+1} \\ x'_{j+1} \end{pmatrix} = \begin{pmatrix} \cos\mu_{\text{step}} & \sin\mu_{\text{step}} \\ -\sin\mu_{\text{step}} & \cos\mu_{\text{step}} \end{pmatrix} \left[\begin{pmatrix} x_j \\ x'_j \end{pmatrix} + \begin{pmatrix} 0 \\ \varepsilon\cos\left(\frac{2\pi j}{N_{\text{step}}}\right)x_j^2 \end{pmatrix} \right]. \quad (10.87)$$

For the integration, the linear tune is Q_0 but the actual tune Q depends on the amplitude of the oscillation. Fig. 10.2 shows results with both quarter-integer and eleventh-integer islands. These islands are stable since the the tune is amplitude dependent; there are actually many overlapping resonances. Rather than just a linear growth, the tune shifts as the amplitude would increase, thus reducing the strength of the excited resonance. In effect the growth becomes limited so that the oscillation is stable. For large enough amplitudes, however, the particle will may be outside the separatrix and can be lost.

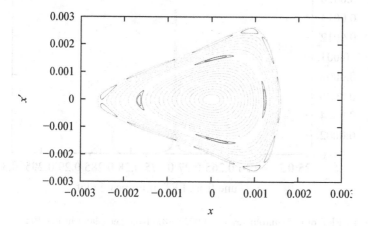

Fig. 10.2 Left: Poincaré plot for 2048 turns and initial x_0 from -0.0002 to -0.003 in steps of -0.0001, and with $N_{step} = 100$, $Q = 0.232$ and initial slope $x'_0 = 0$ and various values for ϵ and the initial x_0 value. The four quarter-integer islands show in the contours with $x_0 = -0.0017$ and -0.0018. The 3/11-integer islands appear with $x_0 = -0.0025$.

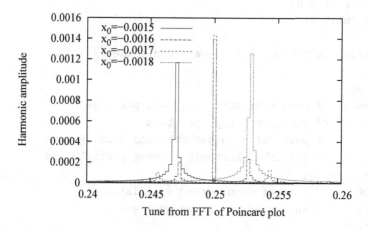

Fig. 10.3 Right: This shows FFT harmonic analysis for the turns with initial $x_0 = -0.0015$, -0.0016, -0.0017, and -0.0018 which bracket the 1/4-integer islands. Notice that the two values in the islands show a strong resonance at 0.25.

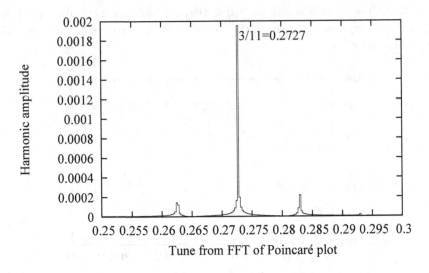

Fig. 10.4   FFT of the contour ($x_0 = -00025$.) 3/11-integer island in Fig 10.2.

10.2.4   *Listing of C program for the simulation*

```
#include <stdio.h>
#include <stdlib.h>
#include <math.h>

// d^2 x/dtheta^2 + Q_0^2 x = eps * cos(m*theta) * x^p

double eps;
int Nturns;    // number of turns for each particle
int Nsteps;    // number of steps per turn
int p;         // power of x in perturbation driving term
int m;         // azimuthal harmonic driving perturbation
double xi0, xpi0;
double Q0;     // tune of unperturbed oscillator
int npart;     // number of particles to track
double rat;

FILE *f;
//=========
int process_cmd(int argc, char** argv)
```

```
{
  int narg, i;

  narg = 10+1;
  if(argc!=narg)
    {
      fprintf(stderr,
      "Usage: %s Nturns, Nsteps, eps, m, pow, x0, x'0, Q0,"
      argv[0]);
      fprintf(stderr, " npart, rat\n");
      fprintf(stderr, " Nturns:\t num turns per particle\n");
      fprintf(stderr, " Nsteps:\t num integ steps/turn\n");
      fprintf(stderr, " eps:\t    epsilon\n");
      fprintf(stderr, " m:\t azimuthal harmonic\n");
      fprintf(stderr, " pow:\t power of x in drive term\n");
      fprintf(stderr, " xi0:\t init x for 1st particle\n");
      fprintf(stderr, " xpi0:\t init x' for 1st particle\n");
      fprintf(stderr, " Q_0 for unperturbed oscillator\n");
      fprintf(stderr, " npart:\t num of particles\n");
      fprintf(stderr, " ratio:\t multiplier of [xi0, xpi0]");
      fprintf(stderr, " for succeeding particles\n");
      return -1;
    }
  i = 1;
  Nturns = atoi(argv[i++]);
  Nsteps = atoi(argv[i++]);
  eps    = atof(argv[i++]);
  m      = atoi(argv[i++]);
  p      = atoi(argv[i++]);
  xi0    = atof(argv[i++]);
  xpi0   = atof(argv[i++]);
  Q0     = atof(argv[i++]);
  npart  = atoi(argv[i++]);
  rat    = atof(argv[i++]);

  return 0;
}
//-------
void dumppar(void)
```

```
{
  fprintf(f, "# %s:\t %d \n", "Nturns", Nturns);
  fprintf(f, "# %s:\t %d \n", "Nsteps", Nsteps);
  fprintf(f, "# %s:\t %f \n", "eps", eps);
  fprintf(f, "# %s:\t %d \n", "m", m);
  fprintf(f, "# %s:\t %d \n", "p", p);
  fprintf(f, "# %s:\t %f \n", "xi0", xi0);
  fprintf(f, "# %s:\t %f \n", "xpi0", xpi0);
  fprintf(f, "# %s:\t %f \n", "Q0", Q0);
  fprintf(f, "# %s:\t %d \n", "npart", npart);
  fprintf(f, "# %s:\t %f \n", "rat", rat);
}
//-------
void trackit(void)
{
  double x, xp, xi, xpi, xt, xtp;
  double fdrive;
  double mustep;
  double ma, mb, mc, md;  // linear matrix
  int nturn, ipart, jstep, kpow;
  double pi;
  double mtheta, dmtheta;
  double a, b, c, d;

  pi = acos(-1.0);
  mustep = 2.0*pi*Q0/((double) Nsteps);
  a = cos(mustep);
  b = sin(mustep);
  c = -b;
  d = a;
  dmtheta = 2.*pi*m/((double) Nsteps);

  xi  = xi0;
  xpi = xpi0;
  for(ipart=0;ipart<npart;ipart++)
    {
      x  = xi;
      xp = xpi;
      fprintf(f, "%6d %6d  %e  %e\n", ipart, 0, x, xp);
```

```
          for(nturn=1;nturn<Nturns;nturn++)
{
   mtheta = 0;
   for(jstep=0;jstep<Nsteps;jstep++)
     {
        xt   = a*x+b*xp;
        xtp = c*x+d*xp;
        fdrive = eps*cos(mtheta)*dmtheta;
        for(kpow=0;kpow<p;kpow++)
{
   fdrive = fdrive*xt;
}
        x   = xt;
        xp = xtp+fdrive;
        mtheta += dmtheta;
      }
   fprintf(f, "%6d %6d  %e  %e\n", ipart, nturn, x, xp);
}
        //       xi  = xi*rat;
        //       xpi = xpi*rat;
        xi  = xi0*(1.0+(rat-1.0)*ipart);
        xpi = xpi0*(1.0+(rat-1.0)*ipart);
      }
}
//-------
int main(int argc, char** argv)
{
   if(process_cmd(argc, argv)) return -1;

   f = fopen("track.dat", "w");
   dumppar();
   trackit();
   fclose(f);

   return 0;
}
```

10.3   Problem 10–3: Sextupole driven third-integer resonance

Write a small computer program that tracks a particle through a periodic cell using a linear transformation.

a) For various initial conditions plot the stroboscopic representation of the particle at the beginning of the periodic cell.

b) Add a thin sextupole kick at the beginning of the cell and study the effect of this nonlinearity near a third-integer resonance for various initial conditions.

Solution:

This is the 1-d version of the Henon map.[32, 43]

For the linear periodic cell we take a transport matrix:

$$\mathbf{M} = \begin{pmatrix} \cos\mu + \alpha\sin\mu & \beta\sin\mu \\ -\gamma\sin\mu & \cos\mu - \alpha\sin\mu \end{pmatrix}, \tag{10.88}$$

with Twiss parameters β, α, and $\gamma = (1+\alpha^2)/\beta$ at the beginning of the cell and a phase advance of μ. We also assume that the superperiodicity $P = 1$, so that the periodic cell is the the same as the one-turn matrix. For our plots, we set $\beta = 1.0$, $\alpha = 0.2$, and tune $Q = 0.6658$. Our program, written in Octave[24], is listed below in § 10.3.1.

a) Fig. 10.5 shows the mundane tracking of the linear case for initial coordinates $x_0 = 0.01$, 0.02, and 0.03 m. See the nice concentric ellipses. Boring, isn't it. We ran the program three times (once for each case) and combined the PostScript[4] output from the program with a little massaging to enlarge and move the fonts and delete the title.

b) At the beginning of each turn, we now make a thin nonlinear kick so that our one-turn algorithm is

$$\begin{pmatrix} x_{n+1} \\ x'_{n+1} \end{pmatrix} = \mathbf{M}^P \left[\begin{pmatrix} x_n \\ x'_n \end{pmatrix} + \begin{pmatrix} 0 \\ \epsilon x_n^2 \end{pmatrix} \right]. \tag{10.89}$$

Fig. 10.6 shows results of tracking 1000 turns with the nonlinear strength $\epsilon = 0.5$ and initial positions $x_0 = 0.01$, 0.02, 0.0216, 0.0218, and 0.03 m. Note that in the stable region the ellipses are distorted to show an increasingly triangular shape as x_0 is increased up to the separatrix. Outside the separatrix, we see that the trajectories are unbounded. Note

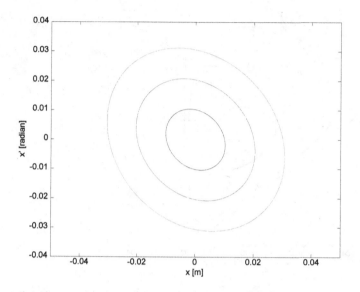

Fig. 10.5 Linear tracking of 1000 turns for initial conditions $x'_0 = 0$ $x_0 = 0.01$, 0.02, and 0.03. The nonlinear coefficient $\epsilon = 0$ in this case.

that the triangular separatrix lies somewhere between the 0.0216 and 0.0218 cases.

Moving the tune closer to 2/3, we see in Fig. 10.7 for an initial position $x = 0.01$ cm, that trajectory is stable with $Q = 0.6662$ but unstable with Q=0.6663. This shows how a beam may be slowly extracted by moving the tune closer the $Q = 2/3$ resonance.

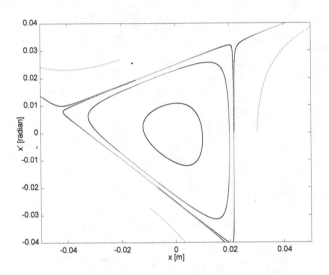

Fig. 10.6   Nonlinear tracking of 1000 turns for initial conditions $x'_0 = 0$, $x_0 = 0.01$, 0.02, 0.216, 0.218 and 0.03. The nonlinear coefficient $\epsilon = 0.5$ in this case.

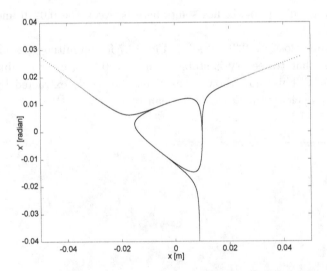

Fig. 10.7   Nonlinear tracking of 1000 turns for initial conditions $x'_0 = 0$, $x_0 = 0.01$. The nonlinear coefficient $\epsilon = 0.5$ in this case. Two cases are shown: a stable case with $Q = 0.6662$ and an unstable case with $Q = 0.6663$.

10.3.1 *Listing of the* Octave *program*

```
#!/usr/bin/octave --persist
% Default parameters:

nturn = 1000;


% Parameters for the periodic cell:
% We will assume a periodicity of 1 for the whole ring:
Qh = 0.6658;
alphah = 0.2;
betah = 1.;


% eps  is the strength of the nonlinear sextupole kick.
eps = 0.5;
% eps = 0.;  % uncomment this line for the linear case.
% Change the following two lines for different initial
% conditions:
x0  = 0.02
xp0 = 0.00


% Define a function to track and plot the result:
function [Qhcheck, x] = trackit(nturn, Qh, betah, alphah,
   eps, x0, xp0)
% Without the sextupole:

  mucell= 2*pi*Qh;

  gammah = (1.+alphah**2)/betah;
  M = [ cos(mucell)+alphah*sin(mucell), betah*sin(mucell);
       -gammah*sin(mucell), cos(mucell)-alphah*sin(mucell)];

  mu = acos(trace(M)/2.);
% if(M(1,2)<0)
%    mu = 2*pi-mu;
% endif
  Qhcheck = mu/2/pi;

  X  = [x0; xp0];
```

```
  j = 1;
  x = [];
  y = [];
 while (j<=nturn)
%    for k=1:1:4
% sextupole at beginning of cell
      XX(1,1) = X(1,1);
      XX(2,1) = X(2,1)+eps*X(1,1)**2;
% track the turn
      X = M*XX;
      x(j) = X(1,1);
      y(j) = X(2,1);
%    endfor;
    j++;
    endwhile

% Plot the data:
    plot(x,  y,  '.');
    str = sprintf("nturn=%d, Q=%6.4f, beta=%3.1f, alpha=%f,\
    eps=%6.3f, x0=%f, xp0=%f",
 nturn, Qh, betah, alphah, eps, x0, xp0);
    title(str, "FontSize", 15)
    axis([-0.05, 0.05, -0.04, 0.04]);
    xlabel("x [m]", "FontSize", 15)
    ylabel("x' [radian]", "FontSize", 15)
% Write out an EPS file of the plot.
% Note that this will overwrite the file "test.eps"
    print -deps test.eps

% Reduce the title's font size to fit in the terminal window.
    title(str, "FontSize", 10);
 endfunction

% Generate the plot with default parameters:
 [Qhcheck, x]=trackit(nturn, Qh, betah, alphah, eps, x0, xp0);

% Write out some help after plotting for the default data:
 printf('Enter "trackit(nturn, Qh, betah, alphah, eps, x0,'
 printf(' xp0)" to track.\n')
```

```
printf('  nturn is number of turns.\n')
printf('  Qh is fractional horizontal tune.\n')
printf('  eps is strength of the sextupole.\n')
printf('  x0 is the initial position in meters.\n')
printf('  xp0 is the initial slope in radians.\n')
printf('Note that this outputs the EPS file called')
printf(' "test.eps" each time')
printf('it is called.\n\n')
printf('\nInitial values are:\n')
```

10.4   Problem 10–4: Synchrotron modulated coupling

By treating the synchrotron oscillations as slowly varying relative to the revolutions, with $p = p_0 + A\sin(Q_s\theta)$, show that a coupling of resonances between the horizontal and vertical betatron oscillations occurs.

Additional note: Perhaps we should have supplied another explicit hint to consider nonlinear terms (see Problems 4-4, 4-5, and 4-6) for this diabolical problem.

Solution:

Let us assume a separated-function lattice with no multipoles other than normal dipoles and quadrupoles.

Recall the Hill's equations from CM: § 5.2 for a ring:

$$x'' + k_x(s)x = \frac{\delta}{\rho(s)}, \qquad \text{(CM: 5.23)}$$

$$y'' + k_y(s)y = 0, \qquad \text{(CM: 5.24)}$$

with the lowest-order periodic "spring constants",

$$k_x(s) = \frac{1}{\rho^2} + \frac{q}{p_0}\frac{\partial B_y}{\partial x}, \qquad \text{(CM: 5.25)}$$

$$k_y(s) = -\frac{q}{p_0}\frac{\partial B_y}{\partial x}. \qquad \text{(CM: 5.26)}$$

Let us add a subscript zero to the first-order k_x and k_y to signify the guide field for the on-momentum particle. Now replace, in the modified Eqs. (CM: 5.25 and CM: 5.26), the design momentum p_0 by p to obtain

$$k_x = \frac{1}{\rho^2} + \frac{q}{p_0(1+\delta)}\frac{\partial B_y}{\partial x} = \frac{1}{\rho^2} + \frac{k_{x0}(s) - \rho^{-2}}{1+\delta}$$

$$= \left(k_{x0}(s) + \frac{\delta}{\rho^2}\right)(1 - \delta + \delta^2 - \cdots)$$

$$= k_{x0} - \left(k_{x0} - \frac{1}{\rho^2}\right)(\delta - \delta^2 + \cdots), \qquad (10.90)$$

$$k_y = -\frac{q}{p_0}\frac{\partial B_y}{\partial x}(1+\delta)^{-1} = k_{y0} - k_{y0}(\delta - \delta^2 + \cdots), \qquad (10.91)$$

with $\delta = (A/p_0)\sin(Q_s\theta)$. Substituting back into Eqs (CM: 5.23 and CM: 5.24) and keeping higher order terms yields

$$x'' + k_{x0}x = \frac{\delta}{\rho} + \left(k_{x0} - \frac{1}{\rho^2}\right)x\delta(1 - \delta + \delta^2 - \cdots), \tag{10.92}$$

$$y'' + k_{y0}y = k_{y0}\, y\delta(1 - \delta + \delta^2 - \cdots), \tag{10.93}$$

In fact, we must be a bit careful, since the δ/ρ term in Eq. (CM: 5.23) was obtained from only the linear and quadratic terms in the expansion of the Hamiltonian of Eq. (CM: 3.83):

$$\mathcal{H} = \frac{q}{p_0}A_s - \left(1 + \frac{x}{\rho}\right)\left(1 + \frac{1}{2}[2\delta + \delta^2 - (x'^2 + y'^2) + \cdots]\right.$$

$$\left. - \frac{1}{8}(4\delta^2 + \cdots) + \cdots\right) + \frac{\partial F_2}{\partial s}$$

$$= \frac{q}{p_0}A_s - \left(1 + \frac{x}{\rho}\right)\left(1 + \delta - \frac{1}{2}(x'^2 + y'^2) + \cdots\right) + 1 + \delta. \quad \text{(CM: 3.83)}$$

Recall from CM: Chapter 7 the equations:

$$\frac{d\omega}{\omega} = -\frac{d\tau}{\tau} = \frac{d\beta}{\beta} - \frac{dL}{L} = \left(\frac{1}{\gamma^2} - \alpha_p\right)\frac{dp}{p}, \tag{CM: 7.2}$$

$$\frac{dW}{dt} = \frac{qv}{2\pi h}(\sin\phi - \sin\phi_s), \tag{CM: 7.25}$$

$$\frac{d\varphi}{dt} \simeq \frac{\omega_{\mathrm{rf}}^2\eta_{\mathrm{tr}}}{\beta^2 U_s}W, \tag{CM: 7.28}$$

where $\phi = \phi_s + \varphi$ and $\eta_{\mathrm{tr}} = \gamma^{-2} - \alpha_p$, that lead to the small amplitude synchrotron oscillation equation

$$\ddot{\varphi} + \Omega_s^2\varphi \simeq 0. \tag{CM: 7.32}$$

The $dL/L = \alpha_p\delta$ term in Eq. (CM: 7.2) will have higher order terms in x' and y' which couple back into Eq. (CM: 7.32).

Perhaps the easiest way to "show" this coupling is with the simulation of a simple FODO lattice constructed of only drifts, quadrupoles, and sector dipoles. We use the results from Problems 4–4, 4–5, and 4–6 to build nonlinear transport functions for these elements. Our lattice consists of $N_c = 60$ periodic cells of the form

$$\mathbf{M}_{\mathrm{cell}} = \mathbf{L}_{\frac{1}{2}} \circ \mathbf{B} \circ \mathbf{L} \circ \mathbf{Q}_{\mathrm{d}} \circ \mathbf{L} \circ \mathbf{B} \circ \mathbf{L} \circ \mathbf{Q}_{\mathrm{f}} \circ \mathbf{L}_{\frac{1}{2}}, \tag{10.94}$$

where $\mathbf{L}$ and $\mathbf{L}_{\frac{1}{2}}$ are nonlinear drift functions, $\mathbf{B}$ represents the dipole function, and $\mathbf{Q}_{\mathrm{f}}$ and $\mathbf{Q}_{\mathrm{d}}$ are the respective functions for focusing and defocusing quadrupoles. Our Octave[24] code for these functions and the

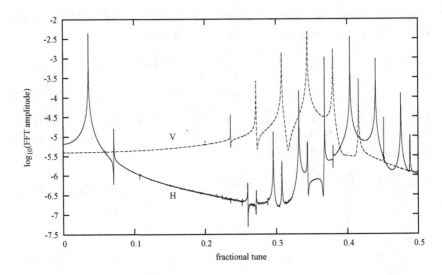

Fig. 10.8   Tune spectra from the horizontal (solid) and vertical (dashed) motion.

tracking code are listed below in §§10.4.1 and 10.4.2. The drift element **L** has a length $l_d = 0.5$ m and the half-length drifts $\mathbf{L}_{\frac{1}{2}}$ at the ends of the periodic cell have length $l_d/2 = 0.25$ m. The two types of quadrupoles have lengths $l_q = 1.0$ m with respective strengths $k_f = 0.2$ m^{-2} and $k_d = -0.202$ m^{-2} for $\mathbf{Q}_f$ and $\mathbf{Q}_d$. The dipoles **B** bend a design particle by an angle $\theta = 3°$ with a radius $\rho = 20$ m. We also take the Lorentz factor to be $\gamma = 100$. (Think of a muon ring, since for protons the required magnetic field would be rather high.) At the end of each turn, we apply a simple linear rf kick using a linear thin-cavity matrix **R** composed of an identity matrix with the rf focusing element modified to be $R_{56} = k_{rf} = 0.004$. Starting with initial vector

$$\mathbf{X}_i = \begin{pmatrix} x_i \\ x_i' \\ y_i \\ y_i' \\ z_i \\ \delta_i \end{pmatrix} = \begin{pmatrix} 0.001 \text{ m} \\ 0 \\ 0.005 \text{ m} \\ 0 \\ 0 \\ 0.002 \end{pmatrix}, \tag{10.95}$$

we track the particle for 4096 turns and calculate the FFT[56] of the x, y and δ to obtain the tune spectrum for each of the three planes.

The Fourier spectra of the transverse planes are plotted in Fig. 10.8. The synchrotron frequency $Q_s = 0.03589$ is quite strong in the horizontal

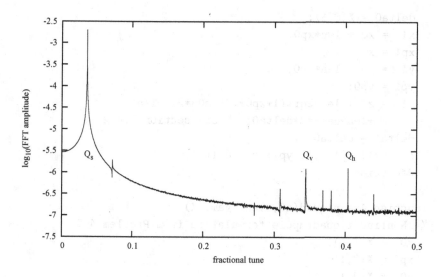

Fig. 10.9   Tune spectrum of the vertical plane.

spectra due to the linear coupling of the δ/ρ term in Eq. (10.92). The maximum betatron tune peaks show up with a lot of synchrotron side bands at $Q_H = 0.4038$ in the horizontal spectrum and $Q_V = 0.3440$ in the vertical spectrum. Notice that many of the vertical peaks also show up as peaks in the horizontal with a much lower amplitude; this illustrates coupling from the vertical oscillations into the horizontal plane. The many sidebands are due to the nonlinear coupling, particularly since the large natural chromaticity from the quadrupoles has not been compensated.

Fig. 10.9 displays the longitudinal spectrum with an obvious peak at the synchrotron tune and smaller peaks at the horizontal and vertical tunes. Here a few synchrotron sidebands are also visible.

10.4.1   *Nonlinear function listings*

```
function A = nldrift(len, gam, X)
%  Nonlinear drift
   x0  = X(1,1);
   xp0 = X(2,1);
   y0  = X(3,1);
   yp0 = X(4,1);
   z0  = X(5,1);
```

```
  delta0 = X(6,1);
  x1  = x0 + len*xp0;
  xp1 = xp0;
  y1  = y0 + len*yp0;
  yp1 = yp0;
  z1  = z0 + len*sqrt(1+xp0**2+yp0**2) -len\
        +len/gam**2*delta0;  % see section 7.4.2
  delta1 = delta0;
  A = [x1; xp1; y1; yp1; z1; delta1];
endfunction;

function A = nlquad(k0, len, gam, X)
%  Nonlinear quadrupole formulation from Problem 4-5
  x0  = X(1,1);
  xp0 = X(2,1);
  y0  = X(3,1);
  yp0 = X(4,1);
  z0  = X(5,1);
  delta0 = X(6,1);
  k = k0/(1+delta0);
  rk = sqrt(k);
  phi = rk*len;
  x1  =  x0 * cos(phi)    + xp0*sin(phi)/rk;
  xp1 = -x0*rk*sin(phi)   + xp0*cos(phi);
  y1  =  y0 * cosh(phi)   + yp0*sinh(phi)/rk;
  yp1 =  y0*rk*sinh(phi)  + yp0*cosh(phi);
  z1  =  z0 + ((x0**2-y0**2)*k+xp0**2+yp0**2)*len/4 \
    + (xp0**2-k*x0**2)*sin(2*phi)/8/rk \
    + (yp0**2+k*y0**2)*sinh(2*phi)/8/rk \
    - x0*xp0*cos(2*phi)/4 \
    + y0*yp0*cosh(2*phi)/4 \
    + len/gam**2*delta0;  % see Section 7.4.2
  delta1 = delta0;
  A = real([x1; xp1; y1; yp1; z1; delta1]);
endfunction;

function A=nlsbend(rho0, phi0, gam, X)
%  Nonlinear sector bend formulation from Problem 4-6
  x0  = X(1,1);
```

```
xp0  = X(2,1);
y0   = X(3,1);
yp0  = X(4,1);
z0   = X(5,1);
delta0 = X(6,1);

rat = sqrt((1+xp0**2)/(1+xp0**2+yp0**2));
theta0 = atan(xp0);
rho1 = rho0*(1+delta0)*rat;
b = (rho0+x0)*cos(phi0)-rho1*cos(phi0+theta0);
c = (rho0+x0)**2-2*(rho0+x0)*rho1*cos(theta0);
ZBp  = (b+sqrt(b**2-c))*sin(phi0);
XBp  = (b+sqrt(b**2-c))*cos(phi0)-rho0;
phi1 = 2*asin( sqrt((XBp-x0)**2+ZBp**2)/2/rho1);
xp1  = tan(theta0+phi0-phi1);
x1   = sqrt(ZBp**2+(XBp+rho0)**2)-rho0;
yp1  = sqrt((1+xp1**2)/(1+xp0**2))*yp0;
y1   = y0 + rho1*phi1*yp0/sqrt(1+xp0**1);
z1   = z0 -(rho1*phi1/rat -rho0*phi0) \
          +rho0*phi0/gam**2*delta0;  % see Section 7.4.2
delta1 = delta0;
A = [x1; xp1; y1; yp1; z1; delta1];
endfunction;
```

10.4.2   Listing of nonlinear FODO lattice simulation

```
#!/usr/bin/env octave

Nt = 1025*4 % number of turns to track
%  Use a FODO lattice of Nc cells.
Nc = 60;
theta = pi/Nc;   % dipole bend design angle
rho = 20.0;      % dipole bend design radius
ld = 0.5;        % design quadrupole length
lq = 1.0;        % design drift length

kf0 =  0.2;
kd0 = -0.202;
krf =  0.004;
```

```
X0 = [0.001; 0.000; 0.005; 0.000; 0.; 0.002];

g0 = 100.0;        % Lorentz factor (gamma)
b0 = sqrt(1.-1./g0**2);
g = g0;

% First, calculate the frequencies for a linear lattice.

L   = drift(ld,g);
Lh  = drift(ld/2., g);
Qf0 = quadrupole(kf0, lq, g);
Qd0 = quadrupole(kd0, lq, g);
B   = sbend(rho, theta, g);
c0  = Lh*B*L*Qd0*L*B*L*Qf0*Lh;    % cell end.
%  Thin rf cavity goes in middle of drift at end of turn.

% Betatron tunes:
Qh0 = Nc*acos((c0(1,1)+c0(2,2))/2.)/2./pi;
Qv0 = Nc*acos((c0(3,3)+c0(4,4))/2.)/2./pi;

Mt0 = c0**Nc;   % one-turn matrix ignoring the rf cavity.
L0 = Nc*2*(ld+rho*theta+ld+lq);  % design circumference
% Use a linear rf cavity of some strength:
Rf = eye(6);
Rf(6,5) = krf

MM = Mt0*Rf;      % Full one-turn matrix with thin rf cavity.

abs(eig(MM))
qvals = arg(eig(MM))/2/pi  % Assume stable in all 3 planes.
Qs0 = qvals(3);            % Might move to a different index.

printf("Qh = %10.6f    Qv = %10.6f    Qs = %10.6f\n",
       Qh0, Qv0, Qs0);

fout = fopen("p10.4.dat", "w");

X = X0;
```

```
g = g0;
% Track the particle through Nt turns
fprintf(fout,
  "%6d  %10.6f  %10.6f  %10.6f  %10.6f  %10.6f  %10.6f\n",
  0, X(1:6,1));
for nt=1:Nt
  kf = kf0/(1+X(6,1));
  kd = kd0/(1+X(6,1));
%% $d\gamma = \gamma\beta^2\delta$, so
  g = g0*(1+b0**2*X(6,1));
  % Track through the nonlinear elements
  for nc = 1:Nc
    X = nldrift(ld/2, g, X);
    X = nlquad(kf, lq, g, X);
    X = nldrift(ld, g, X);
    X = nlsbend(rho, theta, g, X);
    X = nldrift(ld, g, X);
    X = nlquad(kd, lq, g, X);
    X = nldrift(ld, g, X);
    X = nlsbend(rho, theta, g, X);
    X = nldrift(ld/2, g, X);
  endfor;
  X = Rf*X;
  fprintf(fout,
    "%6d  %10.6f  %10.6f  %10.6f  %10.6f  %10.6f  %10.6f\n",\
    nt, X(1:6,1));
%nt, X
endfor;

fclose(fout);
```

Chapter 11

New Problem for Chapter 11: Space-Charge Effects

While the textbook has no problems for Chapter 11, we thought we would add a problem which was suggested by S. Y. Zhang of Brookhaven National Laboratory.

11.1 New Problem 11-1: Beam-beam shift for unequal species

Modify the equation for the beam-beam tune shift, Eq. (CM: 11.34)

$$\Delta Q_{bb} = -\frac{N_{IP} N r_0 \beta_V^*}{2\pi\gamma\sigma_V(\sigma_V + \sigma_H)},$$ (CM: 11.34)

to deal with beams of unequal species, such as for RHIC[61] with fully stripped gold ions ($^{197}\text{Au}^{+79}$) in one ring and deuterons ($^2\text{H}^+$) in the other ring.

Solution:

Recall the formula for the beam-beam tune shift:

$$\Delta Q_{bb} = -\frac{N_{IP} N r_0 \beta_V^*}{2\pi\gamma\sigma_V^2},$$ (CM: 11.33)

where the particle's classical radius is given by

$$r_0 = \frac{q^2}{4\pi\epsilon_0 mc^2},$$ (11.1)

N_{IP} is the number of identical collision points, N is the number of charges per bunch.

Rewriting Eq. (CM: 11.24) for two disparate species, we have for the force on a test charge in beam 1 due to the fields from the distribution of particles in beam 2:

$$F(r) = q_1(E_{r2} + \beta c B_{\theta 2}).$$ (11.2)

The transverse force equation (CM: 11.25) with subscripts added for the two beams becomes

$$F_\perp(y_1) = q_1 \frac{Nq_2}{2\pi\epsilon_0 l}(1 + \beta_2^2)\frac{1 - \exp\left(-\frac{y_1^2}{2\sigma_{V2}^2}\right)}{y_1},$$ (11.3)

for the force on the test particle q_1 due to the fields of beam 2, where the position y_1 is the vertical position of the test particle(1) relative to the center of the "lens" of the opposing beam(2). So we see that the q^2 in the classical radius formula, must be replaced by the product of the particle charges of the two beams, and the mass should be the mass of the test particle, i. e.

$$r_0 \to \frac{q_1 q_2}{4\pi\epsilon_0 m_1 c^2}.$$ (11.4)

The beam distribution parameters in Eq. (CM: 11:34) must be referred to beam 2, so the desired formula for the tune shift of beam 1 should now be

$$\Delta Q_{bb,1} = -\frac{N_{IP}N_2 q_1 q_2 \beta_{V2}^*}{2\pi\epsilon_0 m_1 c^2 \gamma_2 \sigma_{V2}(\sigma_{V2} + \sigma_{H2})},$$ (11.5)

where the numerical subscripts "1" and "2" refer to the various parameters for the respective beams. For the beam-beam tune shift of the other beam, we must just interchange the subscripts 1 and 2.

Chapter 12

Problems of Chapter 12: How to Baffle Liouville

12.1 Problem 12–1: New cooling ideas

Invent a new method of beating Liouville and win the Nobel prize.

Solution:

OK. You've "stumped the chumps"[47, 52].

This is really a suggested problem for further research (in the vein of Knuth's[38] rating "HM50"), so we will not be giving a solution here. As for winning the Nobel prize: Simon van der Meer[67] did it, so why not you?

However, before you come up with your own wacko idea, it is worth mentioning a couple of new ideas which are presently being pursued. Note that our list of references is not all encompassing, but are given as a place to get started.

The first is ionization cooling[26, 18] of a muon beam for use either in a muon collider[6], or for high-brightness source of neutrinos[10, 9]. This uses energy loss from multiple scattering[42] (see § 5.12) of the muon beam through a low density material combined with acceleration through high power rf cavities to reduce the phase-space volume of the beam. The effect is similar to the and radiation damping discussed in CM: §§8.2–4, in that each muon, on average, loses momentum along its direction of motion, but is then accelerated by the cavities along the s-axis.

Another idea is coherent electron cooling of ion beams[22, 45, 69] which combines electron cooling with stochastic cooling. In one straight section the electron beam moves colinearly with the hadron beam so that the hadron beam imprints its density function onto the electron beam. The two beams are separated and a plasma instability (initialized by the previous interaction with the hadron beam) is let to grow, perhaps with enhanced amplification of the distribution through an undulator magnet. Later, the electrons with the amplified signal are run colinearly with the hadron beam in another straight section with the correct phase to reduce the emittance of the hadron beam. It is basically stochastic cooling using a single electron beam as the pickup, amplifier, and kicker.

Chapter 13

Problems of Chapter 13: Spin Dynamics

Note that the **R** matrices in problems 13–6 and 13–7 should be replaced by the $\mathbf{D}^{\frac{1}{2}}$ rotation matrices for spin-$\frac{1}{2}$ as given in Eqs. (CM: 13.182–184).

13.1 Problem 13–1: Translation invariance of spin

In the center-of-mass rest system, show that the integral in Eq. (CM:13.2)

$$\vec{S} = \vec{J} = \int \vec{r} \times [\rho_m(\vec{r})\vec{v}(\vec{r})]\, d^3r, \qquad \text{(CM:13.2)}$$

is invariant if the origin is translated by a fixed vector $\vec{a}$. Assume that the rotation is nonrelativistic and for simplicity that the mass distribution has cylindrical symmetry.

Solution:

First, let us write the position as

$$\vec{x} = \vec{a} + \vec{r}, \qquad (13.1)$$

where the position of the center of mass

$$\vec{a} = \frac{\int \vec{x}\,\rho_m(\vec{x})\,d^3x}{\int \rho(\vec{x})\,d^3x} = \frac{\int \vec{x}\,\rho_m(\vec{x})\,d^3x}{m}. \qquad (13.2)$$

Shifting the origin we can rewrite Eq. (CM: 13.2) as

$$\vec{S} = \int \rho_m(\vec{a} + \vec{r})\,(\vec{a} + \vec{r}) \times \frac{d(\vec{a} + \vec{r})}{dt}\, d^3r, \qquad (13.3)$$

but $\vec{a}$ is a fixed vector, so we have

$$\vec{S} = \int \rho_{\text{cm}}(r)\,(\vec{a} + \vec{r}) \times \frac{d\vec{r}}{dt}\, d^3r, \qquad (13.4)$$

where $\rho_{\rm cm}(r)$ is a cylindrical mass distribution which is symmetric about the center of mass $\vec{a}$.

$$(\vec{a} + \vec{r}) \times \frac{d\vec{r}}{dt} = (\vec{a} + \vec{r}) \times (\dot{r}\hat{r} + r\dot{\theta}\hat{\theta}). \tag{13.5}$$

Since the center of mass of our rigid body is stationary, the individual mass elements rotate with a fixed radius r about the center of mass. So $\dot{r} = 0$, and we get

$$(\vec{a} + \vec{r}) \times \frac{d\vec{r}}{dt} = (\vec{a} + \vec{r}) \times r\dot{\theta}\hat{\theta} = r\dot{\theta}(\vec{a} \times \hat{\theta} + \hat{z}). \tag{13.6}$$

Substituting back into Eq. 13.4

$$\vec{S} = \dot{\theta} \int\!\!\int \rho_{\rm cm}(r) \int (\vec{a} \times \hat{\theta} + \hat{z}) \, d\theta \, r^2 dr \, dz$$

$$= 2\pi\dot{\theta}\hat{z} \int\!\!\int \rho_{\rm cm}(r) \, r^2 dr \, dz, \tag{13.7}$$

since the nonzero part of $\vec{a} \times \hat{\theta}$ will have a $\sin(\theta + \theta_0)$ dependence which integrates to zero. But this is just what we get from Eq. (CM: 13.2) with the cylindrical distribution centered on the origin.

13.2 Problem 13–2: Rapidity: hyperbolic rotation angle

Show that a Lorentz boost by the velocity $\vec{v} = \beta c\, \hat{z}$ has the form

$$\Lambda(\vec{v}) = \begin{pmatrix} \cosh\eta & 0 & 0 & \sinh\eta \\ 0 & 1 & 0 & 0 \\ 0 & 0 & 1 & 0 \\ \sinh\eta & 0 & 0 & \cosh\eta \end{pmatrix}, \tag{13.8}$$

with the hyperbolic rotation angle $\eta = \frac{1}{2}\ln\left(\frac{1+\beta}{1-\beta}\right)$.

Solution:

Let us recall the expression of the hyperbolic functions

$$\cosh\eta = \frac{e^{\eta} + e^{-\eta}}{2}, \quad \text{and} \quad \sinh\eta = \frac{e^{\eta} - e^{-\eta}}{2}, \tag{13.9}$$

with

$$e^{\pm\eta} = \exp\left[\pm\frac{1}{2}\ln\left(\frac{1+\beta}{1-\beta}\right)\right] = \left(\frac{1+\beta}{1-\beta}\right)^{\pm\frac{1}{2}} = \begin{cases} \sqrt{\frac{1+\beta}{1-\beta}}, & \text{for } +, \\ \sqrt{\frac{1-\beta}{1+\beta}}, & \text{for } -. \end{cases} \tag{13.10}$$

Hence we have:

$$2\cosh\eta = \frac{\sqrt{1+\beta}}{\sqrt{1-\beta}} + \frac{\sqrt{1-\beta}}{\sqrt{1+\beta}} = \frac{1+\beta+1-\beta}{\sqrt{1-\beta^2}} = 2\gamma \tag{13.11}$$

and

$$2\sinh\eta = \frac{\sqrt{1+\beta}}{\sqrt{1-\beta}} - \frac{\sqrt{1-\beta}}{\sqrt{1+\beta}} = \frac{1+\beta-1+\beta}{\sqrt{1-\beta^2}} = 2\beta\gamma \tag{13.12}$$

which will yield the well known matrix

$$\Lambda(\vec{v}) = \begin{pmatrix} \gamma & 0 & 0 & \beta\gamma \\ 0 & 1 & 0 & 0 \\ 0 & 0 & 1 & 0 \\ \beta\gamma & 0 & 0 & \gamma \end{pmatrix} \tag{13.13}$$

characterizing the Lorentz boost by a velocity parallel to the z-axis.

Comments:

Note that this hyperbolic rotation "angle" η is also known by the names:[55] *rapidity*, *boost parameter*, and *velocity parameter* with

$$\eta = \frac{1}{2} \ln \left(\frac{U + |\vec{p}|}{U - |\vec{p}|} \right) = \frac{1}{2} \ln \left(\frac{1 + \beta}{1 - \beta} \right). \tag{13.14}$$

We should also note that high energy physicists frequently call[55] *longitudinal rapidity*

$$y = \frac{1}{2} \ln \left(\frac{U + p_\parallel}{U - p_\parallel} \right), \tag{13.15}$$

just rapidity, so there is a possibility of some confusion. Here $p_\parallel$ is the longitudinal momentum component (projected along the beam direction) for an outgoing particle from a collision. In high energy physics there is also a related parameter called *pseudorapidity*[50]

$$\eta = - \ln \left(\frac{\theta}{2} \right), \quad \text{with} \quad y \to \eta \quad \text{as} \quad \beta \to 1, \tag{13.16}$$

where θ is the angle between the outgoing particle from a collision and the direction of the beam. Note that the commonly used symbol η in Eq. (13.16) for pseudorapidity is different from the boost parameter of Eq. (13.14).

13.3   Problem 13–3: Lorentz boost matrix $\Lambda^\mu_\nu(\vec{\beta})$

Show that the matrix for a Lorentz boost[27] by the contravariant proper velocity defined in Eq. (CM:13.26)

$$\begin{pmatrix} \beta^0 \\ \beta^1 \\ \beta^2 \\ \beta^3 \end{pmatrix} = \begin{pmatrix} \gamma \\ \gamma\beta_x \\ \gamma\beta_y \\ \gamma\beta_z \end{pmatrix}. \tag{CM:13.26}$$

may be written in the form given in Eq. (CM:13.28).

$$\Lambda(\vec{\beta})^\mu_{\;\nu}: \quad \Lambda(\vec{\beta}) = \begin{pmatrix} \beta^0 & \beta^1 & \beta^2 & \beta^3 \\ \beta^1 & 1 + \frac{\beta^1\beta^1}{1+\beta^0} & \frac{\beta^1\beta^2}{1+\beta^0} & \frac{\beta^1\beta^3}{1+\beta^0} \\ \beta^2 & \frac{\beta^2\beta^1}{1+\beta^0} & 1 + \frac{\beta^2\beta^2}{1+\beta^0} & \frac{\beta^2\beta^3}{1+\beta^0} \\ \beta^3 & \frac{\beta^3\beta^1}{1+\beta^0} & \frac{\beta^3\beta^2}{1+\beta^0} & 1 + \frac{\beta^3\beta^3}{1+\beta^0} \end{pmatrix}. \tag{CM:13.28}$$

Hint: Start with a boost along the z-axis and apply a rotation of the coordinates to the boost matrix: $\mathbf{R}_x(\zeta)\mathbf{R}_y(\xi)\mathbf{\Lambda}(v\hat{z})\mathbf{R}_y(\xi)^{-1}\mathbf{R}_x(\zeta)^{-1}$.

Solution:

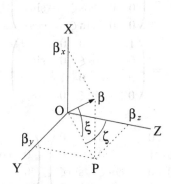

Fig. 13.1   Rotation angles for the rotation from a longitudinal velocity $\beta\hat{z}$ into $\vec{\beta} = \mathbf{R}_x(\zeta)\mathbf{R}_y(\xi)\beta\hat{z}$.

Following the given suggestion, the matrices characterizing the rotations of the axes x and y by respectively an angle ζ and an angle ξ as illustrated in Fig. 13.1, are

$$\mathbf{R}_x(\zeta) = \begin{pmatrix} 1 & 0 & 0 & 0 \\ 0 & 1 & 0 & 0 \\ 0 & 0 & \cos\zeta & \sin\zeta \\ 0 & 0 & -\sin\zeta & \cos\zeta \end{pmatrix}, \quad \text{and} \tag{13.17}$$

$$\mathbf{R}_y(\xi) = \begin{pmatrix} 1 & 0 & 0 & 0 \\ 0 & \cos\xi & 0 & -\sin\xi \\ 0 & 0 & 1 & 0 \\ 0 & \sin\xi & 0 & \cos\xi \end{pmatrix}. \tag{13.18}$$

Applying the spatial rotation $\mathbf{R}_x(\zeta)\mathbf{R}_y(\xi)$ from the velocity 3-vector along $\hat{z}$ to the lab system, we have the transformation

$$\begin{pmatrix} \beta^1 \\ \beta^2 \\ \beta^3 \end{pmatrix} = \begin{pmatrix} \gamma\beta_x \\ \gamma\beta_y \\ \gamma\beta_z \end{pmatrix} = \begin{pmatrix} 1 & 0 & 0 \\ 0 & \cos\zeta & \sin\zeta \\ 0 & -\sin\zeta & \cos\zeta \end{pmatrix} \begin{pmatrix} \cos\xi & 0 & -\sin\xi \\ 0 & 1 & 0 \\ \sin\xi & 0 & \cos\xi \end{pmatrix} \begin{pmatrix} 0 \\ 0 \\ \gamma\beta \end{pmatrix}$$

$$= \begin{pmatrix} 1 & 0 & 0 \\ 0 & \cos\zeta & \sin\zeta \\ 0 & -\sin\zeta & \cos\zeta \end{pmatrix} \begin{pmatrix} -\gamma\beta\sin\xi \\ 0 \\ \gamma\beta\cos\xi \end{pmatrix} = \begin{pmatrix} -\gamma\beta\sin\xi \\ \gamma\beta\sin\zeta\cos\xi \\ \gamma\beta\cos\zeta\cos\xi \end{pmatrix}. \tag{13.19}$$

The inverses of Eqs. (13.17 and 13.18) are respectively

$$\mathbf{R}_x^{-1}(\zeta) = \begin{pmatrix} 1 & 0 & 0 & 0 \\ 0 & 1 & 0 & 0 \\ 0 & 0 & \cos\zeta & -\sin\zeta \\ 0 & 0 & \sin\zeta & \cos\zeta \end{pmatrix}, \quad \text{and} \tag{13.20}$$

$$\mathbf{R}_y^{-1}(\xi) = \begin{pmatrix} 1 & 0 & 0 & 0 \\ 0 & \cos\xi & 0 & \sin\xi \\ 0 & 0 & 1 & 0 \\ 0 & -\sin\xi & 0 & \cos\xi \end{pmatrix}. \tag{13.21}$$

Moreover

$$\mathbf{R}_x(\zeta)\mathbf{R}_y(\xi) = \begin{pmatrix} 1 & 0 & 0 & 0 \\ 0 & 1 & 0 & 0 \\ 0 & 0 & \cos\zeta & \sin\zeta \\ 0 & 0 & -\sin\zeta & \cos\zeta \end{pmatrix} \begin{pmatrix} 1 & 0 & 0 & 0 \\ 0 & \cos\xi & 0 & -\sin\xi \\ 0 & 0 & 1 & 0 \\ 0 & \sin\xi & 0 & \cos\xi \end{pmatrix}$$

$$= \begin{pmatrix} 1 & 0 & 0 & 0 \\ 0 & \cos\xi & 0 & -\sin\xi \\ 0 & \sin\zeta\sin\xi & \cos\zeta & \sin\zeta\cos\xi \\ 0 & \cos\zeta\sin\xi & -\sin\zeta & \cos\zeta\cos\xi \end{pmatrix}, \tag{13.22}$$

and

$$\mathbf{R}_y^{-1}(\xi)\mathbf{R}_x^{-1}(\zeta) = \begin{pmatrix} 1 & 0 & 0 & 0 \\ 0 & \cos\xi & 0 & \sin\xi \\ 0 & 0 & 1 & 0 \\ 0 & -\sin\xi & 0 & \cos\xi \end{pmatrix} \begin{pmatrix} 1 & 0 & 0 & 0 \\ 0 & 1 & 0 & 0 \\ 0 & 0 & \cos\zeta & -\sin\zeta \\ 0 & 0 & \sin\zeta & \cos\zeta \end{pmatrix}$$

$$= \begin{pmatrix} 1 & 0 & 0 & 0 \\ 0 & \cos\xi & \sin\zeta\sin\xi & \cos\zeta\sin\xi \\ 0 & 0 & \cos\zeta & -\sin\zeta \\ 0 & -\sin\xi & \sin\zeta\cos\xi & \cos\zeta\cos\xi \end{pmatrix}.$$

$$(13.23)$$

Planning to carry out the matrix multiplication

$$\mathbf{\Lambda}(\vec{\beta}) = \mathbf{R}_x(\zeta)\mathbf{R}_y(\xi)\mathbf{\Lambda}(v\hat{z})\mathbf{R}_y^{-1}(\xi)\mathbf{R}_x^{-1}(\zeta), \qquad (13.24)$$

with

$$\mathbf{\Lambda}(v\hat{z}) = \begin{pmatrix} \gamma & 0 & 0 & \beta\gamma \\ 0 & 1 & 0 & 0 \\ 0 & 0 & 1 & 0 \\ \beta\gamma & 0 & 0 & \gamma \end{pmatrix}, \qquad (13.25)$$

we proceed step by step, beginning with the product (13.22) × (13.25):

$$\mathbf{R}_x(\zeta)\mathbf{R}_y(\xi)\mathbf{\Lambda}(v\hat{z}) = \begin{pmatrix} 1 & 0 & 0 & 0 \\ 0 & \cos\xi & 0 & -\sin\xi \\ 0 & \sin\zeta\sin\xi & \cos\zeta & \sin\zeta\cos\xi \\ 0 & \cos\zeta\sin\xi & -\sin\zeta & \cos\zeta\cos\xi \end{pmatrix} \begin{pmatrix} \gamma & 0 & 0 & \beta\gamma \\ 0 & 1 & 0 & 0 \\ 0 & 0 & 1 & 0 \\ \beta\gamma & 0 & 0 & \gamma \end{pmatrix}$$

$$= \begin{pmatrix} \gamma & 0 & 0 & \beta\gamma \\ -\beta\gamma\sin\xi & \cos\xi & 0 & -\gamma\sin\xi \\ \beta\gamma\sin\zeta\cos\xi & \sin\zeta\sin\xi & \cos\zeta & \gamma\sin\zeta\cos\xi \\ \beta\gamma\cos\zeta\cos\xi & \cos\zeta\sin\xi & -\sin\zeta & \gamma\cos\zeta\cos\xi \end{pmatrix},$$

$$(13.26)$$

and going on with the product $(13.26) \times (13.23)$:

$$\mathbf{\Lambda}(\vec{\beta}) = \begin{pmatrix} \gamma & 0 & 0 & \beta\gamma \\ -\beta\gamma\sin\xi & \cos\xi & 0 & -\gamma\sin\xi \\ \beta\gamma\sin\zeta\cos\xi & \sin\zeta\sin\xi & \cos\zeta & \gamma\sin\zeta\cos\xi \\ \beta\gamma\cos\zeta\cos\xi & \cos\zeta\sin\xi & -\sin\zeta & \gamma\cos\zeta\cos\xi \end{pmatrix}$$

$$\times \begin{pmatrix} 1 & 0 & 0 & 0 \\ 0 & \cos\xi & \sin\zeta\sin\xi & \cos\zeta\sin\xi \\ 0 & 0 & \cos\zeta & -\sin\zeta \\ 0 & -\sin\xi & \sin\zeta\cos\xi & \cos\zeta\cos\xi \end{pmatrix}$$

$$= \begin{pmatrix} \gamma & -\beta\gamma\sin\xi & \beta\gamma\sin\zeta\cos\xi & \beta\gamma\cos\zeta\cos\xi \\ -\beta\gamma\sin\xi & \Lambda^1{}_1 & \Lambda^1{}_2 & \Lambda^1{}_3 \\ \beta\gamma\sin\zeta\cos\xi & \Lambda^2{}_1 & \Lambda^2{}_2 & \Lambda^2{}_3 \\ \beta\gamma\cos\zeta\cos\xi & \Lambda^3{}_1 & \Lambda^3{}_2 & \Lambda^3{}_3 \end{pmatrix}. \quad (13.27)$$

Recalling Eq. (13.19) while also noting that $(\gamma - 1)(\gamma + 1) = \gamma^2\beta^2$, the terms of $\mathbf{\Lambda}(\vec{\beta})$ evaluate to

$$\Lambda^0{}_0 = \beta^0, \tag{13.28}$$

$$\Lambda^0{}_1 = \Lambda^1{}_0 = \beta^1, \tag{13.29}$$

$$\Lambda^0{}_2 = \Lambda^2{}_0 = \beta^2, \tag{13.30}$$

$$\Lambda^0{}_3 = \Lambda^3{}_0 = \beta^3, \tag{13.31}$$

$$\Lambda^1{}_1 = \cos^2\xi + \gamma\sin^2\xi = 1 + (\gamma - 1)\sin^2\xi = 1 + \frac{(-\beta\gamma\sin\xi)^2}{\gamma + 1}$$

$$= 1 + \frac{\beta^1\beta^1}{1 + \beta^0}, \tag{13.32}$$

$$\Lambda^1{}_2 = \Lambda^2{}_1 = -(\gamma - 1)\sin\zeta\sin\xi\cos\xi = \frac{(-\beta\gamma\sin\xi)(\beta\gamma\sin\zeta\cos\xi)}{\gamma + 1}$$

$$= \frac{\beta^1\beta^2}{1 + \beta^0}, \tag{13.33}$$

$$\Lambda^1{}_3 = \Lambda^3{}_1 = -(\gamma - 1)\cos\zeta\sin\xi\cos\xi = \frac{(-\beta\gamma\sin\xi)(\beta\gamma\cos\zeta\cos\xi)}{\gamma + 1},$$

$$= \frac{\beta^1\beta^3}{1 + \beta^0}, \tag{13.34}$$

$$\Lambda^2{}_2 = \sin^2\zeta\sin^2\xi + \cos^2\zeta + \gamma\sin^2\zeta\cos^2\xi = 1 - (\gamma - 1)\sin^2\zeta\cos^2\xi$$

$$= 1 + \frac{(\beta\gamma\sin\zeta\cos\xi)^2}{\gamma + 1} = 1 + \frac{\beta^2\beta^2}{1 + \beta^0}, \tag{13.35}$$

$$\Lambda^3{}_3 = \cos^2 \zeta \sin^2 \xi + \sin^2 \zeta + \gamma \cos^2 \zeta \cos^2 \xi = 1 - (\gamma - 1) \cos^2 \zeta \cos^2 \xi$$

$$= 1 + \frac{(\beta\gamma \cos \zeta \cos \xi)^2}{\gamma + 1} = 1 + \frac{\beta^3 \beta^3}{1 + \beta^0},$$

$$(13.36)$$

$$\Lambda^2{}_3 = \Lambda^3{}_2 = \sin \zeta \cos \zeta \sin^2 \xi - \sin \zeta \cos \zeta + \gamma \sin \zeta \cos \zeta \cos^2 \xi$$

$$= (\gamma - 1) \sin \zeta \cos \zeta \cos^2 \xi$$

$$= \frac{(\beta\gamma \sin \zeta \cos \xi)(\beta\gamma \cos \zeta \cos \xi)}{\gamma + 1} = \frac{\beta^2 \beta^3}{1 + \beta^0}.$$

$$(13.37)$$

These are just in the desired form of Eq. (CM:13.28).

13.4   Problem 13–4: Thomas precession

Verify that Eqs. (CM: 13.32, CM: 13.35, and CM: 13.39)

$$\begin{pmatrix} \gamma(1+\gamma^2\beta\,\delta\beta\cos\theta) \\ 0 \\ \gamma\,\delta\beta\sin\theta \\ \gamma(\beta+\gamma^2\,\delta\beta\cos\theta) \end{pmatrix} + \mathcal{O}(\delta\beta^2). \tag{CM: 13.32}$$

$$\Delta_1 = \begin{pmatrix} \gamma^3\beta\cos\theta & 0 & -\gamma\sin\theta & -\gamma^3\cos\theta \\ 0 & 0 & 0 & 0 \\ -\gamma\sin\theta & 0 & 0 & \frac{\gamma^2\beta\sin\theta}{\gamma+1} \\ -\gamma^3\cos\theta & 0 & \frac{\gamma^2\beta\sin\theta}{\gamma+1} & \gamma^3\beta\cos\theta \end{pmatrix}. \tag{CM: 13.35}$$

$$\Lambda(-\delta\vec{\beta}') = \begin{pmatrix} 1 & 0 & -\gamma\,\delta\beta\sin\theta & -\gamma^2\,\delta\beta\cos\theta \\ 0 & 1 & 0 & 0 \\ -\gamma\,\delta\beta\sin\theta & 0 & 1 & 0 \\ -\gamma^2\,\delta\beta\cos\theta & 0 & 0 & 1 \end{pmatrix} + \mathcal{O}(\delta\beta^2). \tag{CM: 13.39}$$

are correct to first order in $\delta\beta$.

Solution:

The boost relative to the lab frame Σ_{lab} is

$$t' = \gamma\left(t - \frac{\vec{v}\cdot\vec{r}}{c^2}\right) \implies ct' = \gamma(ct - \vec{\beta}\cdot\vec{r}), \tag{13.38}$$

$$\vec{r}' = \vec{r} - \left[\frac{(\gamma-1)(\vec{v}\cdot\vec{r})}{v^2} - \gamma t\right]\vec{v} = \vec{r} - \left[\frac{(\gamma-1)(\vec{\beta}\cdot\vec{r})}{\beta^2} - \gamma ct\right]\vec{\beta}. \tag{13.39}$$

If a variation $\delta\vec{v}$ is supplied, the velocity and the Lorentz factor become

$$\vec{w} = \vec{v} + \delta\vec{v} = c(\vec{\beta} + \delta\vec{\beta}), \quad \text{and} \tag{13.40}$$

$$\gamma_{\text{w}} \simeq \gamma\left(1 + \gamma^2\frac{\vec{v}\cdot\delta\vec{v}}{c^2}\right) = \gamma(1 + \gamma^2\vec{\beta}\cdot\delta\vec{\beta}), \tag{13.41}$$

due to the following set of approximations:

$$w^2 = \vec{w}\cdot\vec{w} \simeq v^2 + 2\vec{v}\cdot\delta\vec{v} = c^2(\beta^2 + 2\vec{\beta}\cdot\delta\vec{\beta}), \tag{13.42}$$

$$\gamma_{\mathrm{w}} = \left[1 - \frac{w^2}{c^2}\right]^{-\frac{1}{2}} \simeq (1 - \beta^2 - 2\vec{\beta} \cdot \delta\vec{\beta})^{-\frac{1}{2}} = \left[\frac{1}{\gamma^2} - 2\vec{\beta} \cdot \delta\vec{\beta}\right]^{-\frac{1}{2}}, \quad (13.43)$$

$$\gamma_{\mathrm{w}} \simeq \gamma(1 + \gamma^2 \vec{\beta} \cdot \delta\vec{\beta}). \quad (13.44)$$

Let us consider Eq. (CM: 13.30):

$$(\vec{\beta} + \delta\vec{\beta})c = c \begin{pmatrix} 0 \\ \delta\beta \sin\theta \\ \beta + \delta\beta \cos\theta \end{pmatrix}; \quad \text{(CM: 13.30)}$$

the corresponding proper velocity is

$$(\beta'')^{\mu} = \begin{pmatrix} \gamma_{\mathrm{w}} \\ 0 \\ \gamma_{\mathrm{w}} \delta\beta \sin\theta \\ \gamma_{\mathrm{w}}(\beta + \delta\beta \cos\theta) \end{pmatrix} \simeq \begin{pmatrix} \gamma(1 + \gamma^2 \beta \delta\beta \cos\theta) \\ 0 \\ \gamma \delta\beta \sin\theta \\ \gamma(\beta + \gamma^2 \delta\beta \cos\theta) \end{pmatrix}, \quad (13.45)$$

which is just Eq. (CM: 13.32), since

$$\vec{\beta} = \begin{pmatrix} 0 \\ 0 \\ \beta \end{pmatrix}, \quad \text{and} \quad \delta\vec{\beta} = \begin{pmatrix} 0 \\ \delta\beta \sin\theta \\ \delta\beta \cos\theta \end{pmatrix}, \quad (13.46)$$

$$\vec{\beta} \cdot \delta\vec{\beta} = \delta\beta \cos\theta, \quad \text{and} \quad \gamma_{\mathrm{w}} \simeq \gamma(1 + \gamma^2 \beta \delta\beta \cos\theta), \quad (13.47)$$

with

$$\gamma_{\mathrm{w}} \delta\beta \sin\theta = \gamma \delta\beta \sin\theta + \gamma^3 \beta(\delta\beta)^2 \sin\theta \cos\theta \simeq \gamma \delta\beta \sin\theta, \quad (13.48)$$

and with

$$\gamma_{\mathrm{w}}(\beta + \delta\beta \cos\theta) \simeq \gamma(\beta + \delta\beta \cos\theta + \beta^2 \gamma^2 \delta\beta \cos\theta) \quad (13.49)$$

$$\simeq \gamma[\beta + (1 + \beta^2 \gamma^2)\delta\beta \cos\theta] \quad (13.50)$$

$$\simeq \gamma(\beta + \gamma^2 \delta\beta \cos\theta). \quad (13.51)$$

Due to the velocity variation considered so far, the boost of Σ'' relative to the lab system can be written as

$$ct'' = \gamma_{\mathrm{w}}[ct - (\vec{\beta} + \delta\vec{\beta}) \cdot \vec{r}], \quad (13.52)$$

$$\vec{r}'' = \vec{r} + \frac{\gamma_{\mathrm{w}} - 1}{(\vec{\beta} + \delta\vec{\beta})^2}[(\vec{\beta} + \delta\vec{\beta}) \cdot \vec{r} - \gamma_{\mathrm{w}} ct](\vec{\beta} + \delta\vec{\beta}). \quad (13.53)$$

Let us start by working out the time transformation:

$$ct'' \simeq (\gamma + \gamma^3 \vec{\beta} \cdot \delta\vec{\beta})[ct - (\vec{\beta} + \delta\vec{\beta}) \cdot \vec{r}]$$

$$\simeq (\gamma + \gamma^3 \vec{\beta} \cdot \delta\vec{\beta})ct - [\gamma\vec{\beta} + \gamma\delta\vec{\beta} + \gamma^3(\vec{\beta} \cdot \delta\vec{\beta})\vec{\beta}] \cdot \vec{r}$$

$$\simeq (\gamma + \gamma^3 \beta \cos\theta \delta\beta)ct - (\gamma\delta\beta \sin\theta)y - [\beta\gamma + \gamma(1 + \beta^2\gamma^2)\delta\beta \cos\theta]z$$

$$\simeq (\gamma + \gamma^3 \beta \cos\theta \delta\beta)ct - (\gamma\delta\beta \sin\theta)y - (\beta\gamma + \gamma^3 \delta\beta \cos\theta)z, \quad (13.54)$$

having considered that $\vec{\beta} \cdot \vec{r} = \beta z$ and $\delta\vec{\beta} \cdot \vec{r} = (\delta\beta \sin\theta\, y + \delta\beta \cos\theta\, z)$. Then, let us begin to deal with Eq. (13.53) which will be rewritten as

$$\vec{r}'' \simeq \vec{r} + \frac{\beta - 2\delta\beta \cos\theta}{\beta^3}(\gamma_{\mathrm{w}} - 1)[(\vec{\beta} + \delta\vec{\beta}) \cdot \vec{r} - \gamma_{\mathrm{w}} ct](\vec{\beta} + \delta\vec{\beta}), \quad (13.55)$$

having employed

$$(\vec{\beta} + \delta\vec{\beta})^{-2} = (\beta^2 + 2\vec{\beta} \cdot \delta\vec{\beta})^{-1} \simeq (\beta^2 + 2\beta\delta\beta \cos\theta)^{-1}$$

$$\simeq \frac{1}{\beta^2}\left(1 + 2\frac{\delta\beta \cos\theta}{\beta}\right)^{-1}$$

$$\simeq \frac{\beta - 2\delta\beta \cos\theta}{\beta^3}. \quad (13.56)$$

Hence we have:

$$\vec{r}'' \simeq \vec{r} + \frac{\beta - 2\delta\beta \cos\theta}{\beta^3}(\gamma - 1 + \gamma^3\beta\delta\beta \cos\theta)$$

$$[(\vec{\beta} + \delta\vec{\beta}) \cdot \vec{r} - (\gamma + \gamma^3\beta\delta\beta \cos\theta)ct](\vec{\beta} + \delta\vec{\beta}), \quad (13.57)$$

which yields

$$x'' = x, \qquad \text{since} \qquad (\vec{\beta} + \delta\vec{\beta})_x = 0. \quad (13.58)$$

Next we shall have:

$$y'' = y + \left[\frac{\beta - 2\delta\beta \cos\theta}{\beta^3}(\gamma - 1 + \gamma^3\beta\delta\beta \cos\theta)(y\delta\beta \sin\theta + z\delta\beta \cos\theta)\right.$$

$$\left. -(\gamma + \gamma^3\beta\delta\beta \cos\theta)ct\right]\delta\beta \sin\theta$$

$$\simeq y + \frac{\gamma - 1}{\beta}\delta\beta \sin\theta\, z - \gamma \sin\theta\delta\beta\, ct$$

$$\simeq [-\gamma \sin\theta\delta\beta]\, ct + [0]\, x + [1]\, y + \left[\frac{\gamma^2\beta \sin\theta}{\gamma + 1}\delta\beta\right] z,$$

$$(13.59)$$

since

$$\frac{\gamma - 1}{\beta} = \frac{\gamma - 1}{\beta^2\gamma^2}\beta\gamma^2 = \frac{\gamma - 1}{\gamma^2 - 1}\beta\gamma^2 = \frac{\beta\gamma^2}{\gamma + 1}. \quad (13.60)$$

In order to speed up the evaluation of z'' we notice that

$$z'' = z + [\text{the same as for } y''](\beta + \delta\beta \cos\theta)$$

$$= z + [\ldots]\delta\beta \cos\theta + [\ldots]\beta, \quad (13.61)$$

or

$$z'' = z + \frac{\gamma - 1}{\beta}\delta\beta \cos\theta\, z - \gamma\delta\beta \cos\theta\, ct + [\ldots]\beta \quad (13.62)$$

with

$$[\ldots] = \left[\frac{\beta - 2\delta\beta \cos\theta}{\beta^3}(\gamma - 1 + \gamma^3 \beta\delta\beta \cos\theta)(y\delta\beta \sin\theta + z\delta\beta \cos\theta) \right.$$
$$\left. -(\gamma + \gamma^3 \beta\delta\beta \cos\theta)ct \right] \delta\beta \sin\theta\beta. \tag{13.63}$$

Hence we have:

$$z'' - [\gamma\beta + \gamma(1 + \beta^2\gamma^2)]\, ct + \frac{\gamma - 1}{\beta}\delta\beta \sin\theta\, y + C_z\, z, \tag{13.64}$$

where

$$C_z = 1 + \gamma - 1 + \frac{\gamma - 1}{\beta}\delta\beta \cos\theta - \frac{\gamma - 1}{\beta}\delta\beta \cos\theta + \gamma^3\delta\beta \cos\theta$$
$$= \gamma + \gamma^3 \beta\delta\beta \cos\theta, \tag{13.65}$$

and finally

$$z'' = -[\gamma\beta + \gamma^3\delta\beta \cos\theta]\, ct + [0]\, x + \left[\frac{\gamma^2\beta}{\gamma + 1}\delta\beta \cos\theta \right] y$$
$$+ [\gamma + \gamma^3\delta\beta \cos\theta]\, z. \tag{13.66}$$

Therefore, combining Eqs. (13.54, 13.58, 13.59, and 13.66), we obtain:

$$\begin{pmatrix} ct'' \\ x'' \\ y'' \\ z'' \end{pmatrix} = \begin{pmatrix} \gamma + \gamma^3\beta \cos\theta\delta\beta & 0 & -\gamma \sin\theta\delta\beta & -\gamma\beta - \gamma^3 \cos\theta\delta\beta \\ 0 & 1 & 0 & 0 \\ -\gamma \sin\theta\delta\beta & 0 & 1 & \frac{\gamma^2\beta \sin\theta}{\gamma+1}\delta\beta \\ -\gamma\beta - \gamma^3 \cos\theta\delta\beta & 0 & \frac{\gamma^2\beta \sin\theta}{\gamma+1}\delta\beta & \gamma + \gamma^3\beta \cos\theta\delta\beta \end{pmatrix} \begin{pmatrix} ct \\ x \\ y \\ z \end{pmatrix}. \tag{13.67}$$

and the 4×4 matrix M_{tot} can be split into the sum of two matrices

$$M_{\text{tot}} = \begin{pmatrix} \gamma & 0 & 0 & -\gamma\beta \\ 0 & 1 & 0 & 0 \\ 0 & 0 & 1 & 0 \\ -\gamma\beta & 0 & 0 & \gamma \end{pmatrix} + \begin{pmatrix} \gamma^3\beta \cos\theta & 0 & -\gamma \sin\theta & -\gamma^3 \cos\theta \\ 0 & 0 & 0 & 0 \\ -\gamma \sin\theta & 0 & 0 & \frac{\gamma^2\beta \sin\theta}{\gamma+1} \\ -\gamma^3 \cos\theta & 0 & \frac{\gamma^2\beta \sin\theta}{\gamma+1} & \gamma^3\beta \cos\theta \end{pmatrix} \delta\beta, \tag{13.68}$$

where the second matrix on the right is just Δ_1 of Eq. (CM: 13.35), so

$$M_{\text{tot}} = \Lambda(-\vec{\beta}) + \Delta_1\delta\beta = (\text{CM: }13.34) + (\text{CM: }13.35)\,\delta\beta. \tag{13.69}$$

Boost the four-vector of Eq. (CM: 13.32) from Σ_{lab} to Σ':

$$\begin{pmatrix} \gamma & 0 & 0 & -\gamma\beta \\ 0 & 1 & 0 & 0 \\ 0 & 0 & 1 & 0 \\ -\gamma\beta & 0 & 0 & \gamma \end{pmatrix} \begin{pmatrix} \gamma + \gamma^3 \beta\delta\beta \cos\theta \\ 0 \\ \gamma \delta\beta \sin\theta \\ \gamma\beta + \gamma^3 \delta\beta \cos\theta \end{pmatrix} = \begin{pmatrix} 1 \\ 0 \\ \gamma \delta\beta \sin\theta \\ \gamma^2 \delta\beta \cos\theta \end{pmatrix}. \tag{13.70}$$

Since this is an infinitesimal velocity in frame Σ', the corresponding three-vector is simply

$$\delta\vec{\beta}' = \begin{pmatrix} 0 \\ \gamma\,\delta\beta\,\sin\theta \\ \gamma^2\,\delta\beta\,\cos\theta \end{pmatrix}. \tag{13.71}$$

For the corresponding Lorentz boost in the Σ' system, we have

$$\bar{\gamma} = \left\{1 - [\gamma^2\,\delta\beta^2(\sin^2\theta + \gamma^2\cos^2\theta)]\right\}^{-1/2} = 1 + \mathcal{O}(\delta\beta^2), \tag{13.72}$$

so we find to first order in $\delta\beta$:

$$\Lambda(\delta\beta') = \begin{pmatrix} 1 & 0 & \gamma\,\delta\beta\sin\theta & \gamma^2\,\delta\beta\cos\theta \\ 0 & 1 & 0 & 0 \\ \gamma\,\delta\beta\sin\theta & 0 & 1 & 0 \\ \gamma^2\,\delta\beta\cos\theta & 0 & 0 & 1 \end{pmatrix}, \tag{13.73}$$

and so flipping the sign of $\delta\beta'$ we obtain

$$\Lambda(-\delta\beta') = \begin{pmatrix} 1 & 0 & -\gamma\,\delta\beta\sin\theta & -\gamma^2\,\delta\beta\cos\theta \\ 0 & 1 & 0 & 0 \\ -\gamma\,\delta\beta\sin\theta & 0 & 1 & 0 \\ -\gamma^2\,\delta\beta\cos\theta & 0 & 0 & 1 \end{pmatrix}, \tag{13.74}$$

thus verifying Eq. (CM: 13.39).

Correction to Eq. (CM: 13.42):

There were a number of typos in Eq. (CM: 13.42) which we correct here. The equation should have been:

$$\begin{aligned} M^\mu{}_\nu &= \Lambda(-\delta\vec{\beta}')^\mu{}_\kappa \mathbf{R}_x(\theta)^\kappa{}_\nu + \mathcal{O}(\delta\beta'^2) \\ &= \mathbf{R}_x(\theta)^\mu{}_\kappa \Lambda(-\delta\vec{\beta}')^\kappa{}_\nu + \mathcal{O}(\delta\beta'^2). \end{aligned} \tag{13.75}$$

13.5 Problem 13–5: Covariant Thomas-Frenkel-BMT equation

Starting from the covariant form of the Thomas-Frenkel-BMT equation Eq. (CM:13.106) [with typo corrected]

$$\frac{dS^\mu}{d\tau} = \frac{q}{m} \left[F^{\mu\nu} + \frac{g-2}{2} (F^{\mu\nu} + \beta^\mu F^{\nu\kappa} \beta_\kappa) \right] S_\nu, \qquad \text{(CM:13.106)}$$

show that this becomes Eq. (CM:13.69)

$$\frac{d\vec{S}'}{dt} = -\frac{q}{\gamma m} \left[(1 + G\gamma)\vec{B}_\perp + (1 + G)B_\parallel + \gamma \left(G + \frac{1}{\gamma + 1} \right) \frac{\vec{E} \times \vec{v}}{c^2} \right] \times \vec{S}',$$

$$\text{(CM:13.69)}$$

when we express the position, time, and electromagnetic fields in the fixed lab system and spin in the instantaneous rest system of the particle.

Solution:

With $G = (g - 2)/2$ we can write Eq. (CM:13.106) as

$$\frac{dS^\mu}{d\tau} = \frac{q}{m} [(1 + G)F^{\mu\nu}S_\nu + G(S_\nu F^{\nu\kappa}\beta_\kappa)\beta^\mu]. \qquad (13.76)$$

The covariant form of the Lorentz force may be written as

$$\frac{dp^\mu}{d\tau} = qc F^{\mu\nu} \beta_\nu, \qquad (13.77)$$

with

$$F^{\mu\nu}\beta_\nu : \begin{pmatrix} 0 & -E_x/c & -E_y/c & -E_z/c \\ E_x/c & 0 & -B_z & B_y \\ E_y/c & B_z & 0 & -B_x \\ E_z/c & -B_y & B_x & 0 \end{pmatrix} \begin{pmatrix} \gamma \\ -\gamma\beta_x \\ -\gamma\beta_y \\ -\gamma\beta_z \end{pmatrix}$$

$$= \begin{pmatrix} \gamma\vec{E}/c \cdot \vec{\beta} \\ \gamma(E_x/c + B_z\beta_y - B_y\beta_z) \\ \gamma(E_y/c - B_z\beta_x + B_x\beta_z) \\ \gamma(E_z/c + B_y\beta_x - B_x\beta_z) \end{pmatrix}, \qquad (13.78)$$

and we obtain the usual

$$\frac{dp^0}{dt} = \frac{1}{\gamma} \frac{dp^0}{d\tau} = q\,\vec{E} \cdot \vec{\beta}, \qquad (13.79)$$

$$\frac{d\vec{p}}{dt} = \frac{1}{\gamma} \frac{d\vec{p}}{d\tau} = q(\vec{E} + \vec{v} \times \vec{B}). \qquad (13.80)$$

Note that we can write this using a compressed form with

$$F^{\mu\nu} : \begin{pmatrix} 0 & -\vec{E}/c\cdot \\ \vec{E}/c & \vec{B}\times \end{pmatrix} \tag{13.81}$$

for the the field tensor. With this shorthand, we may write

$$qcF^{\mu\nu}\beta_\nu : qc \begin{pmatrix} 0 & -\vec{E}/c\cdot \\ \vec{E}/c & \vec{B}\times \end{pmatrix} \begin{pmatrix} \gamma \\ -\gamma\vec{\beta} \end{pmatrix} = q \begin{pmatrix} \vec{E}\cdot\vec{\beta} \\ \vec{E} + c\vec{\beta}\times\vec{B} \end{pmatrix} \tag{13.82}$$

Evaluation of $F^{\mu\nu}S_\nu$ in the lab system gives:

$$F^{\mu\nu}S_\nu : \begin{pmatrix} 0 & -\vec{E}/c\cdot \\ \vec{E}/c & \vec{B}\times \end{pmatrix} \begin{pmatrix} S^0 \\ -\vec{S} \end{pmatrix} = \begin{pmatrix} \vec{E}\cdot\vec{S}/c \\ S^0\vec{E}/c - \vec{B}\times\vec{S} \end{pmatrix}. \tag{13.83}$$

The scalar term $S_\nu F^{\nu\kappa}\beta_\kappa$ is Lorentz invariant and is more easily evaluated in the rest system, yielding:

$$S_\nu F^{\nu\kappa}\beta_\kappa : (0 \quad -\vec{S}\cdot) \begin{pmatrix} 0 & -\vec{E}'/c\cdot \\ \vec{E}'/c & \vec{B}'\times \end{pmatrix} \begin{pmatrix} 1 \\ 0 \end{pmatrix} = -\frac{\vec{E}'\cdot\vec{S}'}{c}, \tag{13.84}$$

where we may recall from Eq. (CM:13.52) that

$$\vec{E}' = \vec{E}_\parallel + \gamma(\vec{E}_\perp + \vec{v}\times\vec{B}). \tag{13.85}$$

So evaluated in the lab frame, Eq. (13.76) becomes

$$\gamma\frac{d}{dt}\begin{pmatrix} S^0 \\ \vec{S} \end{pmatrix} = \frac{q}{mc}\left[(1+G)\begin{pmatrix} \vec{E}\cdot\vec{S} \\ S^0\vec{E} - c\vec{B}\times\vec{S} \end{pmatrix}\right.$$
$$\left. -G\left(\vec{E}_\parallel + \gamma(\vec{E}_\perp + \vec{v}\times\vec{B})\right)\cdot\vec{S}'\begin{pmatrix} \gamma \\ \gamma\vec{\beta} \end{pmatrix}\right]$$
$$= \frac{q}{mc}\left[(1+G)\begin{pmatrix} \vec{E}\cdot\vec{S} \\ S^0\vec{E} \end{pmatrix} - G\gamma(\vec{E}_\parallel + \gamma\vec{E}_\perp)\cdot\vec{S}'\begin{pmatrix} 1 \\ \vec{\beta} \end{pmatrix}\right.$$
$$\left. -(1+G)\begin{pmatrix} 0 \\ c\vec{B}\times\vec{S} \end{pmatrix} - G\gamma^2(c\vec{B}\times\vec{S}')\cdot\vec{\beta}\begin{pmatrix} 1 \\ \vec{\beta} \end{pmatrix}\right], \tag{13.86}$$

or after separating the temporal and spatial parts:

$$\frac{dS^0}{dt} = \frac{q}{\gamma mc}\left[(1+G)\vec{E}\cdot\vec{S} - G\gamma(\vec{E}_\parallel + \gamma\vec{E}_\perp)\cdot\vec{S}'\right.$$
$$\left. -G\gamma^2(c\vec{B}\times\vec{S}')\cdot\vec{\beta}\right], \tag{13.87}$$

$$\frac{d\vec{S}}{dt} = \frac{q}{\gamma mc}\left[(1+G)S^0\vec{E} - G\gamma(\vec{E}_\parallel + \gamma\vec{E}_\perp)\cdot\vec{S}')\vec{\beta}\right.$$
$$\left. -(1+G)(c\vec{B}\times\vec{S}) - G\gamma^2(\vec{B}\times\vec{S}')\cdot\vec{v}\vec{\beta}\right]. \tag{13.88}$$

The 3-d spin vector transforms from the lab to the rest system as

$$\vec{S}' = \vec{S}_\perp + \frac{1}{\gamma}\vec{S}_\parallel = \vec{S} + \left(\frac{1}{\gamma} - 1\right)\vec{S}_\parallel = \vec{S} + \frac{1-\gamma}{\gamma\beta^2}(\vec{\beta}\cdot\vec{S})\vec{\beta}$$

$$= \vec{S} - \frac{S^0\gamma\vec{\beta}}{\gamma+1}, \tag{13.89}$$

and we also have in the reverse direction

$$\vec{S} = \vec{S}'_\perp + \gamma\vec{S}'_\parallel. \tag{13.90}$$

In order to convert the spin derivative on the left side Eq. (13.88) from $\vec{S}$ to $\vec{S}'$, we must differentiate Eq. (13.89):

$$\begin{aligned}
\frac{d\vec{S}'}{dt} &= \frac{d\vec{S}}{dt} - \frac{dS^0}{dt}\frac{\gamma\vec{\beta}}{\gamma+1} - \frac{S^0}{(\gamma+1)}\frac{d(\gamma\vec{\beta})}{dt} + \frac{S^0\gamma\vec{\beta}}{(\gamma+1)^2}\frac{d\gamma}{dt}\\
&= \frac{d\vec{S}}{dt} - \frac{dS^0}{dt}\frac{\gamma\vec{\beta}}{\gamma+1} - \frac{S^0}{\gamma+1}\frac{q(\vec{E}+\vec{v}\times\vec{B})}{mc} + \frac{S^0\gamma\vec{\beta}}{(\gamma+1)^2}\frac{q\vec{E}\cdot\vec{\beta}}{mc}\\
&= \frac{d\vec{S}}{dt} - \frac{dS^0}{dt}\frac{\gamma\vec{\beta}}{\gamma+1} - \frac{q}{\gamma mc}\frac{S^0}{\gamma+1}\left[\gamma(\vec{E}+\vec{v}\times\vec{B}) - \frac{\gamma^2\beta^2\,E_\parallel}{\gamma+1}\right]\\
&= \frac{d\vec{S}}{dt} - \frac{dS^0}{dt}\frac{\gamma\vec{\beta}}{\gamma+1} - \frac{q}{\gamma mc}\frac{S^0}{\gamma+1}\left[\gamma(\vec{E}+\vec{v}\times\vec{B}) - (\gamma-1)E_\parallel\right]\\
&= \frac{d\vec{S}}{dt} - \frac{dS^0}{dt}\frac{\gamma\vec{\beta}}{\gamma+1} - \frac{q}{\gamma mc}\frac{S^0\vec{E}'}{\gamma+1}.
\end{aligned} \tag{13.91}$$

This was perhaps the most conceptually difficult step, and what remains is a lot of cumbersome algebra to wrestle everything into the desired form.

Multiplying by $\gamma mc/q$ to simplify the algebra a little, the first two terms of the right-hand side of Eq. (13.91) are

$$\frac{\gamma mc}{q}\frac{d\vec{S}}{dt} = (1+G)(S^0\vec{E} - c\vec{B}\times\vec{S}) - G\gamma(\vec{E}'\cdot\vec{S}')\vec{\beta}, \tag{13.92}$$

$$\frac{\gamma mc}{q}\frac{dS^0}{dt}\frac{\gamma\vec{\beta}}{\gamma+1} = \left[(1+G)\vec{E}\cdot\vec{S} - G\gamma\vec{E}'\cdot\vec{S}'\right]\frac{\gamma\vec{\beta}}{\gamma+1}, \tag{13.93}$$

so that we obtain

$$\frac{\gamma mc}{q}\frac{d\vec{S}'}{dt} = (1+G)(S^0\vec{E} - c\vec{B}\times\vec{S}) - G\gamma(\vec{E}'\cdot\vec{S}')\vec{\beta}$$

$$- \left[(1+G)\vec{E}\cdot\vec{S} - G\gamma\vec{E}'\cdot\vec{S}'\right]\frac{\gamma\vec{\beta}}{\gamma+1} - \frac{S^0\vec{E}'}{\gamma+1}$$

$$= (1+G)\left(S^0\vec{E} - c\vec{B}\times\vec{S} - \frac{\gamma\vec{E}\cdot\vec{S}}{\gamma+1}\vec{\beta}\right)$$

$$- G\gamma(\vec{E}'\cdot\vec{S}')\vec{\beta} + G\gamma\vec{E}'\cdot\vec{S}'\frac{\gamma\vec{\beta}}{\gamma+1} - \frac{S^0\vec{E}'}{\gamma+1}$$

$$= (1+G)\left(S^0\vec{E} - c\vec{B}\times\vec{S} - \frac{\gamma\vec{E}\cdot\vec{S}}{\gamma+1}\vec{\beta}\right) - \frac{G\gamma(\vec{E}'\cdot\vec{S}')\vec{\beta}}{\gamma+1} - \frac{S^0\vec{E}'}{\gamma+1}$$

$$= \xi_B + \xi_E, \tag{13.94}$$

where we will separate the electric and magnetic terms into ξ_E and ξ_B, respectively.

The magnetic part becomes, with the help of Eqs. (13.85 and 13.90),

$$\frac{\xi_B}{c} = -(1+G)\vec{B}\times\left[\vec{S}' + (\gamma-1)\vec{S}'_{\parallel}\right] - \frac{G\gamma^2(\vec{\beta}\times\vec{B})\cdot\vec{S}'\,\vec{\beta}}{\gamma+1}$$

$$- \frac{(\gamma\vec{\beta}\cdot\vec{S}')(\gamma\vec{\beta}\times\vec{B})}{\gamma+1}$$

$$= -(1+G)\vec{B}\times\vec{S}' - (1+G)(\gamma-1)\vec{B}\times\vec{S}'_{\parallel} - G(\gamma-1)\overrightarrow{(\vec{B}\times\vec{S}')}_{\parallel}$$

$$+ (\gamma-1)\vec{B}\times\vec{S}'_{\parallel}$$

$$= -(1+G)\vec{B}\times\vec{S}' - G(\gamma-1)\vec{B}\times\vec{S}'_{\parallel} - G(\gamma-1)\overrightarrow{(\vec{B}\times\vec{S}')}_{\parallel}$$

$$= -(1+G)\vec{B}\times\vec{S}' - G(\gamma-1)\left[\vec{B}\times\vec{S}'_{\parallel} + \overrightarrow{(\vec{B}\times\vec{S}')}_{\parallel}\right], \tag{13.95}$$

where we have written the parallel component of $\vec{B}\times\vec{S}'$ as

$$\overrightarrow{(\vec{B}\times\vec{S}')}_{\parallel} = [(\vec{B}\times\vec{S}')\cdot\hat{\beta}]\hat{\beta} = \vec{B}\times\vec{S}' - \overrightarrow{(\vec{B}\times\vec{S}')}_{\perp}$$

$$= \vec{B}\times\vec{S}' - \hat{\beta}\times\left[(\vec{B}\times\vec{S}')\times\hat{\beta}\right]$$

$$= \vec{B}\times\vec{S}' - \hat{\beta}\times\left[(\hat{\beta}\cdot\vec{B})\vec{S}' - (\hat{\beta}\cdot\vec{S}')\vec{B}\right]$$

$$= \vec{B}\times\vec{S}' - \vec{B}_{\parallel}\times\vec{S}' + \vec{S}'_{\parallel}\times\vec{B}, \tag{13.96}$$

or on rearranging

$$\vec{B}\times\vec{S}'_{\parallel} + \overrightarrow{(\vec{B}\times\vec{S}')}_{\parallel} = \vec{B}\times\vec{S}' - \vec{B}_{\parallel}\times\vec{S}', \tag{13.97}$$

since the perpendicular component of any vector $\vec{V}$ may be written as $\hat{\beta} \times (\vec{V} \times \hat{\beta})$. Substituting back into the expression for ξ_B/c gives the desired form for the magnetic field:

$$\frac{\xi_B}{c} = -(1+G)\vec{B} \times \vec{S}' - G(\gamma - 1)\left[\vec{B} \times \vec{S}' - \vec{B}_\parallel \times \vec{S}'\right]$$

$$= -(1+G\gamma)\vec{B} \times \vec{S}' + G(\gamma - 1)\vec{B}_\parallel \times \vec{S}'$$

$$= -\left[(1+G)\vec{B}_\parallel + (1+G\gamma)\vec{B}_\perp\right] \times \vec{S}'. \tag{13.98}$$

With the help of the vector identity $\vec{A} \times (\vec{B} \times \vec{C}) = (\vec{A} \cdot \vec{C})\vec{B} - (\vec{A} \cdot \vec{B})\vec{C}$, the unruly electric field part ξ_E simplifies to

$$\xi_E = (1+G)\left[\gamma\vec{\beta} \cdot \vec{S}'\,\vec{E} - \frac{\gamma\vec{E} \cdot [\vec{S}' + (\gamma - 1)\vec{S}'_\parallel]}{\gamma + 1}\,\vec{\beta}\right]$$

$$- \frac{\gamma\beta}{\gamma + 1}\left[G(\vec{E}_\parallel + \gamma\vec{E}_\perp) \cdot \vec{S}'\,\hat{\beta} + S'_\parallel(\vec{E}_\parallel + \gamma\vec{E}_\perp)\right]$$

$$= \frac{\gamma\beta}{\gamma + 1}\left\{(1+G)\left[(\gamma + 1)S'_\parallel\vec{E} - (\vec{E} \cdot \vec{S}')\hat{\beta} - (\gamma - 1)S'_\parallel\vec{E}_\parallel\right]\right.$$

$$\left. - \left[(1+G)S'_\parallel\vec{E}_\parallel + G\gamma\vec{E}_\perp \cdot \vec{S}'\,\hat{\beta} + \gamma S'_\parallel\vec{E}_\perp\right]\right\}$$

$$= \frac{\gamma\beta}{\gamma + 1}\left\{(1+G)S'_\parallel\left[(\gamma + 1)\vec{E} - (\gamma - 1)\vec{E}_\parallel - \vec{E}_\parallel\right]\right.$$

$$\left. - \left[(1+G)\vec{E} \cdot \vec{S}' + G\gamma(\vec{E}_\perp \cdot \vec{S}')\right]\hat{\beta} - \gamma S'_\parallel\vec{E}_\perp\right\}$$

$$= \frac{\gamma\beta}{\gamma + 1}\left\{(1+G)(\hat{\beta} \cdot \vec{S}')(\vec{E} + \gamma\vec{E}_\perp) - (\vec{E} \cdot \vec{S}')\hat{\beta}\right.$$

$$\left. - G\left[(\vec{E} + \gamma\vec{E}_\perp) \cdot \vec{S}'\right]\hat{\beta} - \gamma(\hat{\beta} \cdot \vec{S}')\vec{E}_\perp\right\}$$

$$= \frac{\gamma\beta}{\gamma + 1}\left\{(1+G)\left[(\hat{\beta} \cdot \vec{S}')\vec{E} - (\vec{E} \cdot \vec{S}')\hat{\beta}\right]\right.$$

$$\left. + G\gamma\left[(\hat{\beta} \cdot \vec{S}')\vec{E}_\perp - (\vec{E}_\perp \cdot \vec{S}')\hat{\beta}\right]\right\}$$

$$= \frac{\gamma\beta}{\gamma + 1}\left[(1+G)\vec{S}' \times (\vec{E} \times \hat{\beta}) + G\gamma\vec{S}' \times (\vec{E}_\perp \times \hat{\beta})\right]$$

$$= -\gamma\left(G + \frac{1}{\gamma + 1}\right)(\vec{E} \times \vec{\beta}) \times \vec{S}'. \tag{13.99}$$

Finally, we see that substituting ξ_B and ξ_E into

$$\frac{d\vec{S}'}{dt} = \frac{q}{\gamma m}\frac{\xi_B + \xi_E}{c} \tag{13.100}$$

brings us immediately to Eq. (CM:13.69).

13.6 Problem 13–6: Spin tune with $2\frac{1}{2}$ snakes

Consider a flat ring with snakes and a polarized proton beam.

a) Prove that $\mathbf{R}_{\hat{y}}(\theta)\mathbf{R}_{\hat{a}}(\pi)\mathbf{R}_{\hat{y}}(\theta) = \mathbf{R}_{\hat{a}}(\pi)$, where $\hat{a} = \hat{z}\cos\phi + \hat{x}\sin\phi$, i. e. $\hat{a} \perp \hat{y}$.

b) Find the closed-orbit spin tune for a ring with $2\frac{1}{2}$ snakes with the full-turn spin rotation matrix:

$$\mathbf{M} = \mathbf{R}_{\hat{z}}(\pi/2)\mathbf{R}_{\hat{y}}[G\gamma(\pi - \alpha)]\mathbf{R}_{\hat{x}}(\pi)\mathbf{R}_{\hat{y}}(G\gamma\pi)\mathbf{R}_{\hat{z}}(\pi)\mathbf{R}_{\hat{y}}(G\gamma\alpha). \quad (13.101)$$

c) What is the stable spin direction $\hat{n}_0$ for the matrix in part b?

Solution:

As mentioned at the beginning of the chapter, the $\mathbf{R}_{\hat{n}}$ are really the $\mathbf{D}_{\hat{n}}^{\frac{1}{2}}$ matrices of CM: § 13.7.

a) To simplify the algebra we can write

$$\mathbf{D}_{\hat{y}}^{\frac{1}{2}}(\theta) = \begin{pmatrix} b & 0 \\ 0 & b^* \end{pmatrix}, \quad (13.102)$$

where $b = \cos\theta + i\sin\theta$ and

$$\mathbf{D}_{\hat{a}}^{\frac{1}{2}}(\pi) = \begin{pmatrix} 0 & c \\ -c^* & 0 \end{pmatrix}, \quad (13.103)$$

where $c = \cos\phi + i\sin\phi$. Multiplying the matrices yields

$$\mathbf{D}_{\hat{y}}^{\frac{1}{2}}(\theta)\mathbf{D}_{\hat{a}}^{\frac{1}{2}}(\pi)\mathbf{D}_{\hat{y}}^{\frac{1}{2}}(\theta) = \begin{pmatrix} b & 0 \\ 0 & b^* \end{pmatrix}\begin{pmatrix} 0 & c \\ -c^* & 0 \end{pmatrix}\begin{pmatrix} b & 0 \\ 0 & b^* \end{pmatrix} = \begin{pmatrix} 0 & bc \\ -b^*c^* & 0 \end{pmatrix}\begin{pmatrix} b & 0 \\ 0 & b^* \end{pmatrix}$$

$$= \begin{pmatrix} 0 & bb^*c \\ -bb^*c^* & 0 \end{pmatrix} = \begin{pmatrix} 0 & c \\ -c^* & 0 \end{pmatrix} = \mathbf{D}_{\hat{a}}^{\frac{1}{2}}(\pi),$$

$$(13.104)$$

since $bb^* = 1$.

b) Substituting the $\mathbf{D}_{j}^{\frac{1}{2}}$ matrices into $\mathbf{M}$ and applying the relation from part a) gives

$$\mathbf{M} = \mathbf{D}_{\hat{z}}^{\frac{1}{2}}(\pi/2)\mathbf{D}_{\hat{y}}^{\frac{1}{2}}[G\gamma(\pi - \alpha)]\mathbf{D}_{\hat{x}}^{\frac{1}{2}}(\pi)\mathbf{D}_{\hat{y}}^{\frac{1}{2}}(G\gamma\pi)\mathbf{D}_{\hat{z}}^{\frac{1}{2}}(\pi)\mathbf{D}_{\hat{y}}^{\frac{1}{2}}(G\gamma\alpha)$$

$$= \mathbf{D}_{\hat{z}}^{\frac{1}{2}}(\pi/2)\mathbf{D}_{\hat{y}}^{\frac{1}{2}}(-G\gamma\alpha)\left[\mathbf{D}_{\hat{y}}^{\frac{1}{2}}(G\gamma\pi)\mathbf{D}_{\hat{x}}^{\frac{1}{2}}(\pi)\mathbf{D}_{\hat{y}}^{\frac{1}{2}}(G\gamma\pi)\right]\mathbf{D}_{\hat{z}}^{\frac{1}{2}}(\pi)\mathbf{D}_{\hat{y}}^{\frac{1}{2}}(G\gamma\alpha)$$

$$= \mathbf{D}_{\hat{z}}^{\frac{1}{2}}(\pi/2)\mathbf{D}_{\hat{y}}^{\frac{1}{2}}(-G\gamma\alpha)\mathbf{D}_{\hat{x}}^{\frac{1}{2}}(\pi)\mathbf{D}_{\hat{z}}^{\frac{1}{2}}(\pi)\mathbf{D}_{\hat{y}}^{\frac{1}{2}}(G\gamma\alpha). \quad (13.105)$$

We should also note that

$$\mathbf{D}^{\frac{1}{2}}_{\hat{x}}(\pi)\mathbf{D}^{\frac{1}{2}}_{\hat{z}}(\pi) = \begin{pmatrix} 0 & -1 \\ 1 & 0 \end{pmatrix}\begin{pmatrix} 0 & i \\ i & 0 \end{pmatrix} = \begin{pmatrix} -i & 0 \\ 0 & i \end{pmatrix} = \mathbf{D}^{\frac{1}{2}}_{\hat{y}}(-\pi), \qquad (13.106)$$

so we have

$$\begin{aligned} \mathbf{M} &= \mathbf{D}^{\frac{1}{2}}_{\hat{z}}(\pi/2)\mathbf{D}^{\frac{1}{2}}_{\hat{y}}(-G\gamma\alpha)\mathbf{D}^{\frac{1}{2}}_{\hat{y}}(-\pi)\mathbf{D}^{\frac{1}{2}}_{\hat{y}}(G\gamma\alpha) = \mathbf{D}^{\frac{1}{2}}_{\hat{z}}(\pi/2)\mathbf{D}^{\frac{1}{2}}_{\hat{y}}(-\pi) \\ &= \frac{\sqrt{2}}{2}\begin{pmatrix} 1 & i \\ i & 1 \end{pmatrix}\begin{pmatrix} -i & 0 \\ 0 & i \end{pmatrix} = \frac{\sqrt{2}}{2}\begin{pmatrix} -i & -1 \\ 1 & i \end{pmatrix}. \end{aligned}$$

$$(13.107)$$

Recalling from Eq. (CM:13.188) that precession angle may be obtained from the trace of the matrix $\mathbf{M}$:

$$\text{tr}(\mathbf{M}) = 2\cos\frac{\mu}{2} = 2\cos\frac{2\pi\nu_{\text{sp}}}{2} = 0, \qquad (13.108)$$

which implies that $\cos(\pi\nu_{\text{sp}}) = 0$ or $\nu_{\text{sp}} = \pm\frac{1}{2}$.

c) To find the stable spin direction $\hat{n}_0$, we recall

$$\mathbf{D}^{\frac{1}{2}}_{\hat{n}}(\theta) = \begin{pmatrix} \cos\frac{\theta}{2} + in_y\sin\frac{\theta}{2} & (n_x + in_z)\sin\frac{\theta}{2} \\ (-n_x + in_z)\sin\frac{\theta}{2} & \cos\frac{\theta}{2} - in_y\sin\frac{\theta}{2} \end{pmatrix}. \qquad (\text{CM:13.188})$$

Substituting $\theta = \pi$ for a spin tune of $\frac{1}{2}$, this becomes

$$\mathbf{D}^{\frac{1}{2}}_{\hat{n}}(\pi) = \begin{pmatrix} in_y & (n_x + in_z) \\ -n_x + in_z & -in_y \end{pmatrix}. \qquad (13.109)$$

Comparing with Eq. 13.107, we must have

$$n_x = -\frac{\sqrt{2}}{2}, \qquad (13.110)$$

$$n_y = -\frac{\sqrt{2}}{2}, \qquad (13.111)$$

$$n_z = 0. \qquad (13.112)$$

Of course a vector pointing in the opposite direction:

$$\hat{n} = \begin{pmatrix} \frac{\sqrt{2}}{2} \\ \frac{\sqrt{2}}{2} \\ 0 \end{pmatrix} \qquad (13.113)$$

is also an acceptable solution. With the extra half-snake, the stable spin direction is tipped away from the vertical by $45°$.

Comment:

More generally if we replace the first matrix $\mathbf{R}_{\hat{z}}(\pi/2) = \mathbf{D}_{\hat{z}}^{\frac{1}{2}}(\pi/2)$ in Eq. (13.101) by the more general, rotation $\mathbf{D}_{\hat{b}}^{\frac{1}{2}}(\zeta)$ with a rotation axis $\hat{b} = \hat{z}\cos\beta + \hat{x}\sin\beta$ still in the horizontal plane, the spin tune will still remain $\pm\frac{1}{2}$, and the stable spin direction will tip away from the vertical by an angle $\zeta/2$. Additionally, we may point out that the two full snakes on opposite sides of the ring do not have to have their rotation axes aligned along the longitudinal and vertical as stated in the problem, but must just have them aligned at right angles to each other. For example, the snakes in RHIC are aligned with rotation angles of $\pm 45°$ to the radial axis[5].

In one of the RHIC rings, one of the authors (MacKay) once used a single rotator (configured as a $D_{\hat{x}}^{\frac{1}{2}}(\zeta)$) in conjunction with the pair snakes to tilt the polarization away from the vertical by $45°$. After injecting polarized protons into one ring, the rotator was turned on and ramped from zero field to the half-snake value without loss of polarization.

13.7 Problem 13–7: Spin tune with two snakes

Consider a proton ring with two snakes on opposite sides having a one-turn rotation map for the closed orbit as

$$\mathbf{M} = \mathbf{R}_{\hat{y}}(G\gamma\pi)\mathbf{R}_{\hat{a}}(\pi - \delta)\mathbf{R}_{\hat{y}}(G\gamma\pi)\mathbf{R}_{\hat{z}}(\pi), \qquad (13.114)$$

with

$$\hat{a} = \hat{x}\cos\phi + \hat{z}\sin\phi. \qquad (13.115)$$

Find the spin tune ν_{sp} as a function of δ and ϕ. Plot the ν_{sp} for small values of ϕ and δ as the snake is detuned from the optimum settings.

Solution:

First we must realize that spin-$\frac{1}{2}$ rotation matrices $\mathbf{R}_{\hat{n}}(\theta) = \mathbf{D}_{\hat{n}}^{\frac{1}{2}}(\theta)$ as stated at the beginning of the this chapter.

Since we just want the spin tune, we can permute the matrices to evaluate the 1-turn spin rotation matrix just after the second snake:

$$\begin{aligned}
\mathbf{M}' &= \left[\mathbf{D}_{\hat{y}}^{\frac{1}{2}}(G\gamma\pi)\right]^{-1}\mathbf{M}\mathbf{D}_{\hat{y}}^{\frac{1}{2}}(G\gamma\pi) \\
&= \mathbf{D}_{\hat{a}}^{\frac{1}{2}}(\pi - \delta)\mathbf{D}_{\hat{y}}^{\frac{1}{2}}(G\gamma\pi)\mathbf{D}_{\hat{z}}^{\frac{1}{2}}(\pi)\mathbf{D}_{\hat{y}}^{\frac{1}{2}}(G\gamma\pi) \\
&= \mathbf{D}_{\hat{a}}^{\frac{1}{2}}(\pi - \delta)\mathbf{D}_{\hat{z}}^{\frac{1}{2}}(\pi), \qquad (13.116)
\end{aligned}$$

where we have applied the result from the first part of the previous problem (13–6 a).

$$\begin{aligned}
\mathbf{M}' &= \left[\mathbf{I}\cos\left(\frac{\pi - \delta}{2}\right) + i(\sigma_x\cos\phi + \sigma_z\sin\phi)\sin\left(\frac{\pi - \delta}{2}\right)\right]i\sigma_z \\
&= i\sigma_z\sin\delta - (\sigma_x\sigma_z\cos\phi + \sigma_z^2\sin\phi)\cos\delta \\
&= i\sigma_z\sin\delta - (-i\sigma_y\cos\phi + \mathbf{I}\sin\phi)\cos\delta, \qquad (13.117)
\end{aligned}$$

since $\sigma_x\sigma_z = -i\sigma_y$ and $\sigma_z^2 = \mathbf{I}$.

$$\cos(\pi\nu_{\mathrm{sp}}) = \frac{1}{2}\mathrm{tr}(\mathbf{M}') = -\sin\phi\cos\delta \qquad (13.118)$$

since all three Pauli matrices have trace zero.

For the ideal case, $\phi = \delta = 0$, so $\cos(2\pi\nu_{\rm sp}) = 0$ which gives $\nu_{\rm sp0} = \pm\frac{1}{2}$ for the spin tune. If we write $\nu_{\rm sp} = \frac{1}{2} + \Delta\nu_{\rm sp}$, then

$$\cos(\pi\nu_{\rm sp}) = \cos\left(\frac{\pi}{2} + \pi\Delta\nu_{\rm sp}\right) = -\sin(\pi\Delta\nu_{\rm sp}) = -\sin\phi\cos\delta, \quad (13.119)$$

or on solving for the tune:

$$\nu_{\rm sp} = \frac{1}{2} + \frac{1}{\pi}\sin^{-1}\left(\sin\phi\cos\delta\right). \quad (13.120)$$

For small deviations from the ideal case, Eq. (13.119) expands to

$$\Delta\nu_{\rm sp} \simeq \frac{\phi}{\pi}\left(1 - \frac{\delta^2}{2}\right). \quad (13.121)$$

The leading term is linear in ϕ, and the quadratic dependence on δ only appears when the rotation axes of the two snakes are not perpendicular, i. e. $\phi \neq 0$. The dependence on ϕ and δ in Eq. (13.120) is illustrated in Figs. 13.2 and 13.3, respectively.

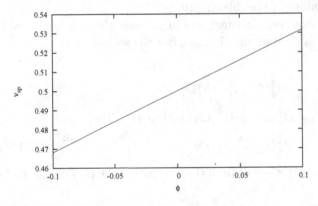

Fig. 13.2 Small-angle linear dependence of the spin tune on snake axis angle ϕ from Eq. (13.120).

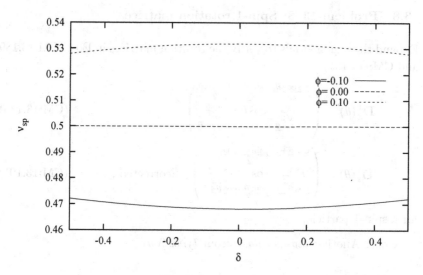

Fig. 13.3   Plot of δ-dependence of Eq. (13.120) showing an approximate quadratic dependence on the deviation of the second snake's rotation angle.

13.8 Problem 13–8: Spin-1 rotation matrices

Expand the exponentials to verify the rotation matrices in Eqs. (CM:13.189 and CM: 13.191)

$$
\mathbf{D}^1_{\hat{x}}(\theta) = \begin{pmatrix} \frac{1+\cos\theta}{2} & \frac{\sin\theta}{\sqrt{2}} & \frac{1-\cos\theta}{2} \\ -\frac{\sin\theta}{\sqrt{2}} & \cos\theta & \frac{\sin\theta}{\sqrt{2}} \\ \frac{1-\cos\theta}{2} & -\frac{\sin\theta}{\sqrt{2}} & \frac{1+\cos\theta}{2} \end{pmatrix}, \tag{CM:13.189}
$$

$$
\mathbf{D}^1_{\hat{z}}(\theta) = \begin{pmatrix} \frac{\cos\theta+1}{2} & i\frac{\sin\theta}{\sqrt{2}} & \frac{\cos\theta-1}{2} \\ i\frac{\sin\theta}{\sqrt{2}} & \cos\theta & i\frac{\sin\theta}{\sqrt{2}} \\ \frac{\cos\theta-1}{2} & i\frac{\sin\theta}{\sqrt{2}} & \frac{\cos\theta+1}{2} \end{pmatrix} \quad \text{[corrected]}, \tag{CM:13.191}
$$

for a spin-1 particle.

Oops: Another *lapsus calami* from *Typos Я us.*

Solution:

Recall Eqs. (CM:13.176 and CM:13.178):

$$
S_x = \frac{1}{\sqrt{2}} \begin{pmatrix} 0 & -i & 0 \\ i & 0 & -i \\ 0 & i & 0 \end{pmatrix}, \tag{CM:176}
$$

$$
S_z = \frac{1}{\sqrt{2}} \begin{pmatrix} 0 & 1 & 0 \\ 1 & 0 & 1 \\ 0 & 1 & 0 \end{pmatrix}. \tag{CM:178}
$$

The squares of S_x and S_z are

$$
S_x^2 = \frac{1}{2} \begin{pmatrix} 1 & 0 & -1 \\ 0 & 2 & 0 \\ -1 & 0 & 1 \end{pmatrix}, \quad \text{and} \quad S_z^2 = \frac{1}{2} \begin{pmatrix} 1 & 0 & 1 \\ 0 & 2 & 0 \\ 1 & 0 & 1 \end{pmatrix}. \tag{13.122}
$$

The third power of S_x and S_z are respectively just

$$
S_x^3 = S_x, \quad \text{and} \quad S_y^3 = S_y. \tag{13.123}
$$

So we may write the spin-1 rotation matrices as

$$
\mathbf{D}^1_j(\theta) = e^{i\theta S_j}
$$

$$
= \mathbf{I} + S_j \left[i\theta + \frac{(i\theta)^3}{3!} + \cdots \right] + S_j^2 \left[\frac{(i\theta)^2}{2!} + \frac{(i\theta)^4}{4!} + \cdots \right]
$$

$$
= \mathbf{I} + iS_j \sin\theta + S_j^2 (\cos\theta - 1), \tag{13.124}
$$

for $\hat{j} \in \{\hat{x}, \hat{z}\}$. (Note that this is true for $\hat{j} = \hat{y}$ as well.) So the spin-1 rotation matrix about the $\hat{x}$ direction is

$$
\mathbf{D}_{\hat{x}}^1(\theta) = \begin{pmatrix} 1 & 0 & 0 \\ 0 & 1 & 0 \\ 0 & 0 & 1 \end{pmatrix} + \frac{\sin\theta}{\sqrt{2}} \begin{pmatrix} 0 & 1 & 0 \\ -1 & 0 & 1 \\ 0 & -1 & 0 \end{pmatrix}
$$

$$
+ \frac{1}{2} \begin{pmatrix} \cos\theta - 1 & 0 & 1 - \cos\theta \\ 0 & 2(\cos\theta - 1) & 0 \\ 1 - \cos\theta & 0 & \cos\theta - 1 \end{pmatrix}
$$

$$
= \begin{pmatrix} \frac{1+\cos\theta}{2} & \frac{\sin\theta}{\sqrt{2}} & \frac{1-\cos\theta}{2} \\ -\frac{\sin\theta}{\sqrt{2}} & \cos\theta & \frac{\sin\theta}{\sqrt{2}} \\ \frac{1-\cos\theta}{2} & -\frac{\sin\theta}{\sqrt{2}} & \frac{1+\cos\theta}{2} \end{pmatrix}. \tag{13.125}
$$

Similarly we have

$$
\mathbf{D}_{\hat{z}}^1(\theta) = \begin{pmatrix} 1 & 0 & 0 \\ 0 & 1 & 0 \\ 0 & 0 & 1 \end{pmatrix} + \frac{i\sin\theta}{\sqrt{2}} \begin{pmatrix} 0 & 1 & 0 \\ 1 & 0 & 1 \\ 0 & 1 & 0 \end{pmatrix}
$$

$$
+ \frac{1}{2} \begin{pmatrix} \cos\theta - 1 & 0 & \cos\theta - 1 \\ 0 & 2(\cos\theta - 1) & 0 \\ \cos\theta - 1 & 0 & \cos\theta - 1 \end{pmatrix}
$$

$$
= \begin{pmatrix} \frac{\cos\theta+1}{2} & \frac{i\sin\theta}{\sqrt{2}} & \frac{\cos\theta-1}{2} \\ \frac{i\sin\theta}{\sqrt{2}} & \cos\theta & \frac{i\sin\theta}{\sqrt{2}} \\ \frac{\cos\theta-1}{2} & \frac{i\sin\theta}{\sqrt{2}} & \frac{\cos\theta+1}{2} \end{pmatrix}, \tag{13.126}
$$

Chapter 14

Problems of Chapter 14: Position Measurements and Spectra

14.1 Problem 14–1: Signal from a ring pickup

A relativistic ($\beta \simeq 1$) bunched beam travels down the beam pipe shown below with a ring pickup. The radius of the pipe and pickup is $b = 4$ cm, and the length of the pickup is $a = 5$ cm. Ignore longitudinal fields. Plot the shape of a voltage pulse seen on an oscilloscope with a 50 Ω termination. Assume that the coaxial cable is also a 50 Ω cable.

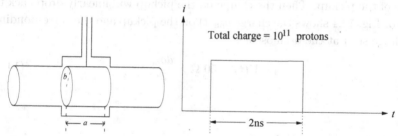

(Note that we have replaced w by a to agree with the figure.)

Solution:

The total charge in the bunch is
$$q = 10^{11} \times 1.6 \times 10^{-19} \text{ C} = 16 \text{ nC}, \tag{14.1}$$
and the bunch length in the lab is
$$l = \beta c \Delta t = 3 \times 10^8 \text{ [m/s]} \times 2 \times 10^{-9} \text{ [s]} = 60 \text{ cm}. \tag{14.2}$$
There must be a total charge of -16 nC on the beam pipe plus pickup, so the maximum charge on the pickup will be
$$\frac{5 \text{ cm}}{60 \text{ cm}} \times (-16 \text{ nC}) = 1.333 \text{ nC}. \tag{14.3}$$

263

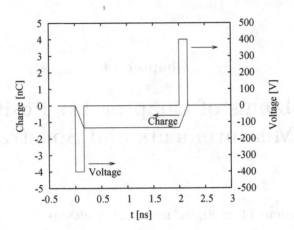

Fig. 14.1 Charge on pickup and voltage across 50 Ω termination.

As the front edge of be beam crosses the upstream edge of the pickup, the charge on the pickup will grow linearly for $1/6$ ns and then remain constant until the trailing edge of the bunch reaches the upstream edge of the of the pickup. Then the charge on the pickup will linearly drop back to zero. Fig. 14.1 shows the charge $q_{\mathrm{pu}}(t)$ on the pickup and the corresponding voltage seen at the oscilloscope

$$V(t) = 50 \ \Omega \times \frac{dq_{\mathrm{pu}}}{dt}. \tag{14.4}$$

14.2 Problem 14–2: Longitudinal Schottky spectrum

Given the following Schottky spectrum for RHIC with $^{197}\text{Au}^{79+}$ ions (fully stripped) where only a single storage cavity was powered, calculate the rf voltage in the cavity. Assume the following parameters for RHIC: $\gamma_{tr} = 22.8$, circumference of $L = 3833.845$ m, harmonic number $h = 7 \times 360 = 2520$, and $U_s = 100$ GeV/nucleon at fixed energy.

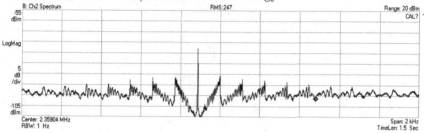

Solution:

Recall the formula for the square of the synchrotron oscillation frequency Eq. (CM: 7.84),

$$\Omega_s^2 = \frac{qVh\eta_{tr}\omega_s^2 \cos\phi_s}{2\pi\beta^2 U_s}, \tag{CM:7.84}$$

with $\eta_{tr} = \gamma^{-2} - \gamma_{tr}^{-2}$ and $\phi_s = 180°$ above transition. We should note that the U_s in this equation is also the total energy of the ion: $U_s = 197 \times 100$ GeV $= 19700$ GeV.

Solving for V we have

$$V = \frac{2\pi\beta^2 U_s}{qh|\eta_{tr}|} \left(\frac{f_{sy}}{f_{rev}}\right)^2, \tag{14.5}$$

where we have written $f_{sy} = \Omega/2\pi$ and $f_{rev} = \omega_s/2\pi$. Counting the number of synchrotron sidebands in 6 divisions we estimate about 10.5 sidebands or about $\frac{6}{10} \times 2$ kHz/10.5 $\simeq 115$ Hz for the synchrotron frequency. Fully stripped gold ions have a mass of 0.93113 GeV per nucleon, although for our purposes, we can just use the mass of a proton 0.93827 GeV. This gives $\gamma \simeq 107$ and $\beta \simeq 0.99996 \simeq 1$, and $|\eta_{tr}| \simeq 0.00184$. The revolution frequency is around 78.2 kHz. Plugging in all these values we find $V \simeq 723$kV.

Bibliography

[1] (2010a). URL www.ubuntu.com.

[2] (2010b). Maxima, a computer algebra system, URL http://maxima.sourceforge.net.

[3] Abramowitz, M. and Stegun, I. A. (eds.) (1970). *Handbook of Mathematical Functions with Formulas, Graphs, and Mathematical Tables* (Dover Publ. Co., New York).

[4] Adobe Systems Incorporated (1990). *PostScript Language Reference Manual, 2nd Ed.* (Addison Wesley).

[5] Alekseev, I. (2003). "Polarized proton collider at RHIC", *Nucl. Inst. and Meth.* **A 499**, p. 392.

[6] Ankenbrandt, C. M. *et al.* (1999). Status of muon collider research and development and future plans, .

[7] Arfkin, G. (1970). *Mathematical Methods for Physicists* (Academic Press, New York).

[8] Bajdich, M., Mitas, L. and Wagner, L. K. (2008). "Pfaffian pairing and back-flow wavefunctions for electronic structure quantum Monte Carlo methods", *Phys. Rev. B* **77**, p. 115112.

[9] Berg, J. S. *et al.* (2006). Cost-effective design for a neutrino factory, *Phys. Rev. STAB* **9**.

[10] Blondel, A. *et al.* (2004). ECFA/CERN Studies of a European Neutrino Factory Complex, .

[11] Brioschi, F. (1856). "Sur l'analogie entre une classe de déterminants d'order pair; et sur les déterminants binaires", *J. für die reine und ang. Math.* **52**, p. 113.

[12] Bryant, P. J. and Johnsen, K. (1993). *The Principles of Circular Accelerators and Storage Rings* (Cambridge University Press, Cambridge).

[13] Cayley, A. (1849). Sur les déterminants gauches, *Journal für die reine und angewandte Mathematik* **38**, p. 93.

[14] Chao, A. *et al.* (1988). "Experimental Investigation of Nonlinear Dynamics in the Fermilab Tevatron", *Phys. Rev. Lett.* **61**, p. 2752.

[15] Chassman, R., Green, G. K. and Rowe, E. M. (1975). Preliminary Design of a Dedicated Synchrotron Radiation Facility, *IEEE Trans. on Nucl. Sci.* **22**,

p. 1765.

[16] Clark, A. (1971). *Elements of Abstract Algebra* (Wadsworth Publ. Co., Belmont, CA).

[17] Cole, F. T. (1969). "Nonlinear Tranformations in Action-Angle Variables", Tech. rep., National Accelerator Lab.

[18] Coney, L. (2009). Status of the MICE Muon Ionization Cooling Experiment, in *Proc. of PAC09* (Vancourver, BC, Canada), p. 1680.

[19] Conte, M. and MacKay, W. W. (1991). *An Introduction to the Physics of Particle Accelerators*, 1st edn. (World Sci., Singapore).

[20] Conte, M. and MacKay, W. W. (2008). *An Introduction to the Physics of Particle Accelerators*, 2nd edn. (World Sci., Singapore).

[21] Courant, E. D. and Snyder, H. S. (1958). "Theory of Alternating Gradient Synchrotron", *Ann. Phys.* **3**, p. 1.

[22] Derbenev, Y. (2007). Use of an electron beam for stochastic cooling, in *Proc. of COOL 2007* (Bad Kreuznach, Germany), p. 149.

[23] Dragt, A. (2011). *Lie Methods for Nonlinear Dynamics with Applications to Accelerator Physics* (University of Maryland), URL http://www.physics.umd.edu/dsat/.

[24] Eaton, J. W. *et al.* (2011). Octave, URL http://www.gnu.org/software/octave.

[25] Edwards, D. A. and Teng, L. C. (1973). "Parametrization of linear coupled motion in periodic systems", in *IEEE Trans. on Nucl. Sci.*, Vol. NS-20, No. 3, p. 885.

[26] Fernow, R. C. and Gallardo, J. C. (1995). Muon transverse ionization cooling: stochastic approach, *Phys. Rev. E* **52**, p. 1039.

[27] Gasiorowicz, S. (1966). *Elementary Particle Physics* (John Wiley and Sons, New York).

[28] Gilmore, R. (1974). *Lie Groups, Lie Algebras, and Some of Their Applications* (John Wiley & Sons, New York).

[29] Goldstein, H. (1980). *Classical Mechanics, 2nd Ed.* (Addison-Wesley, Reading, MA).

[30] Groening, M. and Castellaneta, D. (1987). "The Simpsons", .

[31] Healy, L. M. (1986). *Lie Algebraic Methods for Treating Lattice Parameter Errors in Particle Accelerators*, Ph.D. thesis, University of Maryland, College Park, MD.

[32] Henon, M. and Heiles, C. (1964). "The Applicability of the Third Integral of Motion: Some Numerical Experiments", *Astron. J.* **69**, p. 73.

[33] Iselin, F. C. (1985). "Lie Transformations and Transport Equations for Combined-Function Dipoles", *Particle Accelerators* **17**, p. 143.

[34] Jackson, J. D. (1962). *Classical Electrodynamics* (John Wiley and Sons, New York).

[35] Jacobi, C. G. J. (1866). *Vorlesungen über Dynamik, nebst funf hinterlassenen abhandlungen desselben, herausgegeben von A. Clebsch* (G. Reimer, Berlin).

[36] Jenkins, F. A. and White, H. E. (1976). *Fundamentals of Optics, 4th Ed.* (McGraw-Hill Book Co., New York).

[37] Kernighan, B. W. and Ritchie, D. M. (1988). *The C Programming Language*,

2nd edn. (Prentice Hall, Englewood Cliffs, NJ).

[38] Knuth, D. E. (1973). *The Art of Computer Programming*, Vol. 1 (Addison-Wesley).

[39] Knuth, D. E. (1995). "Overlapping Pfaffians", Electron. J. Combin. 3 (1996), no. 2, #R5.

[40] Lee, S. Y. (2004). *Accelerator Physics* (World Sci.).

[41] Lee, S. Y. et al. (1991). "Experimental Determination of a Nonlinear Hamiltonian in a Synchrotron", *Phys. Rev. Lett.* **67**, p. 3768.

[42] Leo, W. R. (1987). *Techniques for Nuclear and Particle Physics* (Springer-Verlag, Berlin).

[43] Lichtenberg, A. J. and Lieberman, M. A. (1983). *Regular and Stochastic Motion* (Springer-Verlag).

[44] Liouville, J. (1838). "Sur la Théorie de la Variation des constantes arbitraires", *Journal de Math. Pures et Appl.* **3**, p. 342.

[45] Litvinenko, V. N. and Derbenev, Y. S. (2009). "coherent electron cooling", *Phys. Rev. Lett.* **102**, p. 114801.

[46] MacKay, W. W. (2006). "Comment on Healy's Symplectification Algorithm", in *Proceedings of EPAC 2006* (Edinburgh), p. 2284.

[47] Magliozzi, T. and Magliozzi, R. (2009). Car Talk, URL http://www.cartalk.com.

[48] Mane, S. R. (1992). "Solutions of Laplace's equation in two dimensinos with a curved longitudinal axis", *Nucl. Inst. and Meth.* **A321**, p. 365.

[49] Morse, P. M. and Feshbach, H. (1953). *Methods of theoretical physics* (McGraw-Hill).

[50] Nakamura, K. and others (Particle Data Group) (2010). "Review of Particle Physics", *J. Phys. G: Nucl. Part. Phys.* **37**, p. 075021.

[51] Nolte, D. D. (2010). "The tangled tale of phase space", *Physics Today* **63**, April, p. 33.

[52] NPR (2010). National Public Radio, URL http://www.npr.org/.

[53] Ohnuma, S. (1973). "quarter integer resonance by sextupoles", in *Proc. of the U.S.-Japan Seminar on High Energy Accelerator Science* (Tokyo and Tsukuba, Japan), p. 114.

[54] Penrose, R. (2007). *The Road to Reality: a Complete Guide to the Laws of the Universe* (Vintage Books).

[55] Perl, M. L. (1974). *High Energy Hadron Physics* (John Wiley & Sons, New York, NY).

[56] Press, W. H. et al. (1986). *Numerical Recipes* (Cambridge University Press, New York).

[57] Robinson, K. W. (1958). "Radiation Effects in Circular Electron Accelerators", *Phys. Rev.* **111**, p. 373.

[58] Ruggiero, A. G. (1972). "Stochasiticity Limit in One-Dimension", Tech. rep., National Accelerator Lab.

[59] Sands, M. (1970). The physics of electron storage rings an introductions, Tech. Rep. SLAC-121, UC-28 (ACC), University of California, Santa Cruz.

[60] Satogata, T. et al. (1992). "driven respons of a trapped particle beam", *Phys. Rev. Lett.* **68**, p. 1838.

[61] Satogata, T. *et al.* (2003). "Commissioning of RHIC Deuteron-Gold Collision", in *Proc. of PAC2003* (Portland, OR), p. 1706.

[62] Siegel, W. (2002). *Fields* (C. N. Yang Institute for Theoretical Physics, SUNYSB, Stoney Brook, NY), URL http://insti.physics.sunsb.edu/~siegel/plan.html.

[63] Spencer, E. (1987). *The Fairie Queene* (Penguin Books).

[64] Steffen, K. G. (1965). *High Energy Beam Optics* (John Wiley and Sons, New York).

[65] Suzuki, T. (1982). "Hamiltonian Formulation for Synchrotron Oscillations and Sacherer's Integral Equation", *Particle Accelerators* **12**, p. 237.

[66] Tsai, Y.-S. (1974). "Pair production and bremsstrahlung of charged leptons", *Rev. Mod. Phys.* **46**, p. 815.

[67] van der Meer, S. (1984). Stochastic cooling and the accumulation of antiprotons, Nobel Lecture.

[68] Wang, Y. *et al.* (1994). "effects of tune modulation on particles trapped in one-dimensional resonance islands", *Phys. Rev. E* **49**, p. 5697.

[69] Webb, S. D. *et al.* (2011). Effects of e-beam parameters on coherent electron cooling, in *Proc. of 2011 Particle Accelerator Conf.* (New York), p. 232.

[70] Williams, T., Kelly, C. *et al.* (2008). gnuplot, URL http://www.gnuplot.info.

[71] Wilson, M. N. (1983). *Superconducting Magnets* (Clarendon Press, Oxford).

Index